Human Rights and Scientific and
Technological Development

Human Rights and Scientific and
Technological Development

Note to the Reader from the UNU

The UN Commission on Human Rights invited the United Nations University in 1986 to study both the positive and negative impacts of scientific and technological development on human rights and fundamental freedoms. The University responded to the invitation by launching a research project on the interrelationship between human rights and the advances in science and technology, focusing especially on the interaction between socio-cultural, economic, and political factors on the one hand and scientific and technological advances on the other. One aim of the project was to formulate policy recommendations that will strengthen the positive impacts of technological development, and an effort was made to survey the opinions and perceptions of international experts and human rights lawyers in developing countries in order that those recommendations would carry not only social scientific significance but have legal applicability as well.

The present volume deals with the technological implications for development and human rights in Asia and Latin America, the norm-setting role of the United Nations, and the responsibility of the scientific community for promoting human rights. It concludes that science and technology are forces that are now too powerful and too full of ramifications for them to be left to *laissez-faire* attitudes which permit them to take what direction they please. There is an urgent need to guide and channel technology so as to make it an instrument for the furtherance of human rights, particularly in the developing world.

Human Rights and Scientific and Technological Development

Studies on the affirmative use of science and technology for the furtherance of human rights, commissioned as a special project by the United Nations University, following a reference to the University by the United Nations Human Rights Commission

Edited by C.G. Weeramantry

 UNITED NATIONS UNIVERSITY PRESS

The views expressed in this publication are those of the authors and do not necessarily represent the views of the United Nations University.

United Nations University Press
The United Nations University, Toho Seimei Building, 15-1 Shibuya 2-chome, Shibuya-ku, Tokyo 150, Japan   Tel.: (03) 499-2811   Fax: (03)499-2828
Telex: J25442   Cable: UNATUNIV TOKYO

Typeset by Asco Trade Typesetting Limited, Hong Kong
Printed by Permanent Typesetting and Printing Co. Ltd., Hong Kong
Cover design by Tsuneo Taniuchi

HRSTD-1/UNUP-731
ISBN 92-808-0731-5
United Nations Sales No. E.90.III.A.3
03500 C

Contents

Foreword

The present volume constitutes a first response by an interdisciplinary and cross-cultural group of experts under United Nations University auspices to the invitation by the UN Commission on Human Rights to study both the positive and negative impacts of scientific and technological development on human rights and fundamental freedoms. The interrelationship between scientific and technological advances and human rights has attracted in recent years the increasing attention of academics, policy-makers, and the general public. Studies have also been conducted by the United Nations system on the subject, including the publication *Human Rights and Scientific and Technological Development* (UN, 1982). As this publication makes clear, there still exist important lacunae in the understanding of the complex and intricate interface between scientific and technological progress and human rights, not to speak of the practical application of such progress and knowledge to enhance human rights. This book, it is hoped, will make a contribution in this respect.

We are grateful to all the contributors to this volume, and particularly to Professor Christopher G. Weeramantry, who kindly agreed to co-ordinate the research work and edit this book. We were greatly saddened by the passing away of Dr Yo Kubota, who not only played a key role in the project from the UN Centre for Human Rights in Geneva, but was also a conscientious scholar and activist for the course of human rights. He was killed in an accident while working as a member of the UN Transition Assistance Group in Namibia in June 1989.

We wish to express our appreciation to the Ministry of Foreign Affairs of the Government of Japan for its encouragement and financial support for this research project. The research and this publication have been made possible by its generous grant.

Roland Fuchs
Vice-Rector
The United Nations University

Introduction: United Nations Approaches to Human Rights and Scientific and Technological Developments

SADAKO OGATA

The human rights standards established since the founding of the United Nations have become widely recognized in the world today. Starting with the adoption of the Universal Declaration of Human Rights in 1948, tbe United Nations succeeded in enacting the two Covenants on Economic, Social and Cultural Rights and on Civil and Political Rights, as well as a host of more specialized conventions on the Prevention and Punishment of the Crime of Genocide, on the Elimination of All Forms of Racial Discrimination, against Torture or Other Cruel, Inhuman or Degrading Treatment or Punishment, etc. Much of the norm-creating work was carried out in the first 25 years of the United Nations. The main efforts during this period were devoted to the promotion of human rights.

The International Year for Human Rights in 1968 marked a watershed. At the International Conference on Human Rights which was held in Tehran to commemorate the year, delegates assembled to review the progress that had been made in the field of human rights and to prepare the programme for subsequent years. The conference observed that "since the adoption of the Universal Declaration of Human Rights the United Nations has made substantial progress in defining standards for the enjoyment and protection of human rights and fundamental freedoms. During this period many important international instruments were adopted but much remains to be done in regard to the implementation of those rights and freedoms."[1] It became the declared intention of the conference to direct future efforts towards the implementation of the norms that had been established. In other words, the protection more than the promotion of human rights was to become the main focus of human rights activities in the next decades.

The International Conference on Human Rights recommended to the Economic and Social Council that the Commission on Human Rights be requested to prepare model rules of procedure for dealing with violations of human rights. Already in 1966, the Economic and Social Council, in response to the initiative taken by the Special Committee on the Situation with regard to the Implementation of the Declaration on the Granting of Independence to Colonial Countries and Peoples, had already invited the Commission to devise measures to halt violations of human rights and fundamental freedoms in reference particularly to apartheid. The recommendation of the International Conference was designed to encourage the United Nations to expand its authority to examine information relevant to gross violations of human rights and fundamental freedoms, undertake investigations, and make necessary recommendations. The Economic and Social Council adopted resolutions 1235 (XLII), 728 F (XXVIII), and 1503 (XLVIII), each of which marked steps through which the Commission on Human Rights and the Sub-Commission were given extended authority to become actively engaged in the protection of human rights. Fact-finding missions undertaken by specially established committees, by working groups and special rapporteurs, gradually became the mainstay of the United Nations activities in human rights fields. The implementation mechanisms in the various human rights treaties, such as the Covenants and the Convention on the Elimination of All Forms of Racial Discrimination, also began to function in the 1970s.

A second significant impact of the Tehran Conference was to recast the human rights issue in the North–South context. The Covenant on Economic, Social, and Cultural Rights had established the basic economic, social, and cultural rights to be promoted within the nation-state. The North–South framework, however, was to shift the main milieu for the promotion and protection of human rights away from the individual nation-states to the international community, which was divided into the economically developed and deprived countries. The International Conference on Human Rights, in Resolution XVII, recognized that it was the collective responsibility of the international community to ensure the attainment of the minimum standard of living necessary for the enjoyment of human rights and fundamental freedoms by all persons throughout the world, and called upon the developed countries "to facilitate transfer of adequate development resources and technology to the developing countries" and "to make at least one per cent of their Gross National Product available as international aid on equitable terms."[2]

A third development of significance that emanated from the International Conference on Human Rights was the introduction of the question of the impact of recent scientific and technological developments on human rights. The Proclamation of Tehran stated that "while recent scientific discoveries and technological advances have opened vast prospects for economic, social, and cultural progress, such developments may nevertheless endanger the rights and freedoms of individuals and will require continuing attention."[3] The Conference considered the need for "thorough and continuous interdisciplinary studies at both the national and the international level" and recommended to the organizations of the United Nations family that they undertake studies of the problems particularly with regard to:

1. Respect for privacy in view of recording techniques.

2. Protection of the human personality and its physical and intellectual integrity in view of the progress in biology, medicine, and biochemistry.
3. The uses of electronics which may affect the rights of the person and the limits which should be placed on its uses in a democratic society.
4. More generally, the balance which should be established between scientific and technological progress and the intellectual, spiritual, cultural, and moral advancement of humanity.[4]

Later in 1968, the General Assembly responded to the initiative taken by the Conference, and adopted a resolution inviting the Secretary-General to undertake a study of the problems of human rights arising from developments in science and technology, and especially with reference to the areas of concern expressed in Tehran.

In view of the great changes brought about by scientific and technological developments on human life and social progress, it might seem even odd that the issue of the impact of science and technology on human rights had not been raised in the United Nations until 1968. There were, however, political reasons connected with both the North–South and East–West confrontations prevalent in the United Nations that tended to exclude the issue from the human rights fora. To begin with, among the developing countries that comprised a majority in the United Nations, there was a general feeling that scientific and technological developments were not matters causing immediate concern. Moreover, in the context of United Nations politics, these countries tended to show reluctance to allow an agenda not in their direct interest to gain priority attention. Among the more developed industrial countries, the issue contained seeds for East–West confrontation. The Western countries were anxious to promote measures to assure respect for privacy and protection of the human personality against the progress in biology, medicine, and biochemistry. At the United Nations, they were eager to challenge the Eastern socialist countries on the grounds that scientific and technological developments were being applied by them for purposes of state control. George Orwell's "Big Brother" was the symbolic reminder of the political consequences of scientific and technological developments when left in the hands of totalitarian governments.

During the period 1971 to 1976, however, a number of substantive reports on scientific and technological developments and human rights were prepared by the Secretary-General and the specialized agencies to be considered by the General Assembly and the Commission on Human Rights. The reports included a study on respect for the privacy of individuals and the integrity and sovereignty of nations in the light of advances in recording and other techniques; a study on the uses of electronics which may affect the rights of the person, such as computerized data systems and electronic communication techniques; a study on the advances in biology, medicine, and biochemistry, such as artificial insemination, psychotropic drugs and chemicals introduced into food production, packaging, and storage; a study on the use of scientific and technological progress to improve the quality of food, housing, and work; a study on the harmful effects of

automation and mechanization of production, the deterioration of the human
environment, and the destructive power of modern weapons. In 1975, the
United Nations convened a meeting of a group of scientists to discuss the
balance which should be established between scientific and technological developments and the intellectual, spiritual, cultural, and moral advancement of
humanity. The results of the conference were incorporated in a report by the
Secretary-General.

Although these reports raised questions of major importance on a wide range
of human rights problems in the contemporary world, neither the General
Assembly nor the Commission on Human Rights became involved in a substantive examination of the issue. Any attempt to become engaged in a standard-
setting work on human rights and scientific and technological developments
would have required both the commitment and the capability of a number of
delegations. Particularly, the lack of leadership on the part of the Western industrial countries at that time in taking a major initiative in response to the questions
presented by these reports was to be costly. The socialist countries, which became increasingly apprehensive of the possibility that these findings might
prompt the United Nations to opt for greater protection of the individual against
the negative effects of science and technology, decided by a bold stroke to shift
the orientation of the entire subject-matter.

At the twenty-ninth General Assembly in 1974, the Soviet Union, together
with the German Democratic Republic, Hungary, Poland, and others, presented a
draft declaration on the use of scientific and technological progress in the interests
of peace and for the benefit of mankind.[5] The following year, the General Assembly proclaimed the "Declaration on the Use of Scientific and Technological Progress in the Interests of Peace and for the Benefit of Mankind." The declaration was adopted, however, without the support of the Western countries, all
of which abstained from voting. The declaration was notable in that it deviated
from the basic approach that had been followed in the United Nations since 1968
on the question of human rights and scientific and technological developments.
The entire text was an enumeration of the obligations imposed upon states rather
than the protection of individuals. The declaration stated that all states were to
ensure that the results of scientific and technological developments were to be
made use of "in the interests of strengthening international peace and security,
freedom and independence, and also for the purpose of the economic and social
development of peoples." All states were to "refrain from any acts involving the
use of scientific and technological achievements for the purposes of violating the
sovereignty and territorial integrity of other states, interfering in their internal
affairs, waging aggressive wars, suppressing national liberation movements or
pursuing a policy of racial discrimination."[6]

During the next ten years, the United Nations' debate on the issue of human
rights and scientific and technological developments was to be characterized by a
strong East–West emphasis. Focusing on the "right to life," the socialist countries embarked on a linkage of human rights with peace and disarmament. The

emphasis now turned on "the urgent need for all possible efforts by the international community to strengthen peace, remove the threat of war, particularly nuclear war, halt the arms race and achieve general and complete disarmament under effective international control, and prevent violations of the principles of the Charter of the United Nations regarding the sovereignty and territorial integrity of states and self-determination of peoples."[7] Through building the above-mentioned conditions, the "right to life" was to be assured. The socialist countries followed up their initiative and called for the implementation of the principles contained in the declaration. The thirty-fifth General Assembly in 1980 invited the Secretary-General as well as the member states to provide information concerning the implementation of the provisions of the declaration.[8] Subsequent resolutions adopted under the agenda item on human rights and scientific and technological development repeated the call to make efforts to strengthen peace and pursue disarmament, and to utilize the results of scientific and technological developments for the promotion and realization of human rights, particularly the "right to life." The Western countries consistently abstained from voting for resolutions in support of the Declaration on the Use of Scientific and Technological Progress in the Interests of Peace and for the Benefit of Mankind.

In 1981, the Commission on Human Rights, in emphasizing the importance of implementing the provisions of the above declaration, requested the Sub-Commission to undertake a study with particular reference to the "right to work" and the "right to development." The discussion over the examination of these rights had caused considerable division within the Commission. The incorporation of the "right to work" and the "right to development" signified an attempt by the socialist countries to win over the support of the developing countries.

In counteracting the socialist drive, the Western approach to the issue of human rights and scientific and technological developments was no less political. At its thirty-third session in 1977, the Commission on Human Rights adopted the resolution proposed by the United Kingdom to request the Sub-Commission to study "with a view to formulating guidelines, if possible, the question of the protection of those detained on the grounds of mental ill-health against treatment that might adversely affect the human personality and its physical and intellectual integrity."[9] The Western countries chose to focus on the human rights of those political dissidents in the Soviet Union who were said to be detained in mental institutions. In 1980, Mrs Erica-Daes was appointed special rapporteur. She was to prepare guidelines relating to procedures for determining whether adequate grounds existed for detaining persons on the grounds of mental ill-health, and principles for the protection of persons suffering from mental disorder.

Although the debates within the United Nations over the question of the protection of the mentally ill or of persons suffering from mental disorder had strong East–West overtones, the substance of the study undertaken by the special rapporteur dealt with problems and principles of universal relevance. Noting that "improved medical and psychotherapeutic technology can in some cases constitute a threat to the physical and intellectual integrity of the individual," the report indicated that "'scientific and technological products, means and methods have already been misused in some States . . . in particular in the treatment of persons detained on grounds of mental ill-health or mental disorder." The report proceeded to propose principles, guidelines and guarantees to cover legal, medical, economic, and social aspects relating to a patient's admission to an institution, detention, treatment, discharge, and rehabilitation. Governments were expected to adapt their laws to the proposed body of principles, guidelines, and guarantees which were considered to serve as the minimum United Nations standard for the protection of the mentally ill.[10] The Daes report had immediate impact in arousing worldwide interest in the question of the protection of the rights of the mentally ill. In some countries, notably Japan, provisions concerning the admission of mentally ill patients were revised to accord more with the standards set by the United Nations.

Another standard-setting exercise undertaken by the Sub-Commission merits special attention. In 1977, the Commission on Human Rights requested the Sub-Commission to engage in a second study on relevant guidelines in the field of computerized personal files.[11] Mr Louis Joinet acted as the special rapporteur. The interest in this particular subject had dated from the International Conference on Human Rights held in Tehran in 1968. A report had been prepared by the Secretary-General in 1973 relating to "respect for the privacy of individuals and the integrity and sovereignty of nations in the light of advances in recording and other techniques." The Joinet report was significant in that it not only pointed out the dangers of computerized files to the preservation of privacy, or to the enjoyment of freedoms, but also recognized that the exercise of some rights, such as the "right to vote," might be greatly facilitated by the use of data processing. The report attempted to provide that "while the use of manual (or *a fortiori* computerized) personal data files entails an obvious risk of violation of the privacy of individuals, there are cases where, on the contrary, the use of such files makes it possible to promote the effective enjoyment of certain human rights." The report recommended for consideration possible options for preparing minimum standards to be established by national and international legislation.[12]

As shown in the studies on the guidelines, principles, and guarantees for protection of persons detained on grounds of mental ill-health or suffering from mental disorder, and on the guidelines for the regulation of computerized personal data files, some standard-setting efforts have borne results in the protection of human rights vis-à-vis scientific and technological developments. However, it cannot be denied that these two studies represent achievements in rather limited spheres, when taking into consideration the vast areas still left untouched. Whether an incremental approach will eventually meet the required objectives in the field of human rights and scientific and technological developments, or whether the time has come to embark on a more general approach, is a question facing those concerned with the promotion of human rights in the United Nations.

A small step in the direction of a general approach was taken in 1983, when the

Commission on Human Rights adopted a resolution inviting all member states and relevant international organizations to submit their views to the Secretary-General "on the most effective ways and means of using the results of scientific and technological developments for the promotion and realization of human rights and fundamental freedoms."[13] What characterized this particular initiative was the underlying conviction that while vigilance must be kept on the negative effects of scientific and technological developments on human rights, due recognition should be given to the benefits that these achievements bring to the promotion and realization of human rights. In the words of the representative of Japan, who introduced the resolution, "the positive and negative effects of science and technology on human rights are two sides of the same coin" which must be "grasped in their total context."[14] The new approach received support within the Commission, since it attempted to break away from the existing United Nations trend to concentrate almost exclusively on the negative effects of scientific and technological developments on human rights. By focusing both on the positive and negative aspects, greater support was expected from a wider range of member states, cutting across East–West and North–South divisions. In fact the countries in the South showed interest in the issue in its new orientation, since they felt encouraged by the prospect that science and technology would be helpful both in accelerating economic and social development as well as in promoting human rights.

Although the Commission was able to establish a general framework to examine scientific and technological developments in their positive and negative aspects, it turned to the United Nations University and other interested academic and research institutions for extensive examination of the question.[15] To begin with, the task of discerning the positive and negative aspects in itself required expertise of an interdisciplinary character. Defining the requirements of scientific and technological policies, while setting up standards for the protection of human rights, posed challenges of a kind that were beyond the competence of an intergovernmental body. The United Nations University Project on Human Rights and Scientific and Technological Development was set up in response to the invitation by the Commission on Human Rights to probe the complex of interlinking problems.

In view of the vast areas of social and economic life affected by developments in science and technology, it would not be easy to set up a central focus through which to examine the positive and negative effects of scientific and technological developments. Nevertheless, issues involving the right to education, health, or environment might be among the best entry points, since they represent widely acclaimed rights in the international community. Advances in science and technology allow for the spread of education through the dissemination of information, promote universal health care, or assure clean air and water. To the extent that access is assured to promote the attainment of these goals, the positive effects of scientific and technological developments are expanded. On the other hand, science and technology also endanger human rights and human personality when electronics intrude into privacy, medicine turns to human experimentation and genetic engineering, or industrial waste destroys the human habitat. In so far as the negative effects of scientific and technological developments are pronounced, protective measures against violations of human rights become the priority consideration. Of particular importance might be the use of scientific and technological means for the monitoring of trends. Some early-warning mechanisms might also be devised.

Attempts to set a general standard in human rights and scientific and technological developments involve continuing exploration. As science and technology develop, the nature of their impact on human rights also changes. It is in this ever-evolving context that the issue must be fully examined. The treatment of the subject-matter within the United Nations has been far from adequate. However, with the importance of the effects of scientific and technological developments on human life and human rights becoming better understood, and with the political confrontations within the United Nations somewhat subsiding, the time may be ripe to make greater efforts to reach agreement on basic principles and standards for the promotion and protection of human rights. The United Nations should take a lead in bringing about substantive international co-operation for issues of global significance.

NOTES

1. United Nations, *Final Act of the International Conference on Human Rights*, Tehran, 22 April to 13 May 1968, p. 4.
2. United Nations (note 1 above), p. 14.
3. United Nations (note 1 above), p. 5.
4. United Nations (note 1 above), p. 12.
5. General Assembly Resolution 3268 (XXIX), 3269 (XXIX), 1974.
6. General Assembly Resolution 3384 (XXX), 1975.
7. General Assembly Resolution 37/189, 1982.
8. General Assembly Resolution 35/130A, 1980.
9. Commission on Human Rights Resolution 10A (XXXIII), 1977.
10. "Draft Body of Principles, Guidelines and Guarantees for the Protection of the Mentally Ill and of Persons Suffering from Mental Disorder," E/CN.4/Sub.2/1985/20, Annex.
11. Commission on Human Rights Resolution 10B (XXXIII), 1977.
12. "Study of the Relevant Guidelines in the Field of Computerized Personnel Files," E/CN.4/Sub.2/1983/18.
13. Commission on Human Rights Resolution 1983/41, 1983.
14. Commission on Human Rights Resolution 1984/27, 1984.
15. Commission on Human Rights Resolution 1986/9, 1986.

Part 1

Scope and Objectives

1

The Problems, the Project, and the Prognosis

C.G. WEERAMANTRY

The impact of science and technology on society has long been acknowledged. That impact can be beneficial or detrimental. The detrimental aspects have attracted considerable comment and analysis, especially in recent years, and these studies have led naturally to a consideration of the adverse impact of science and technology on human rights.

There has been much concentration, in the recent literature, on the ways in which both specific human rights and general human rights principles are being undermined by advances in science and technology. Such concentrated studies of these adverse impacts ought not, however, to distract us from examining the other side of the coin.

On 10 March 1986 the UN Commission on Human Rights adopted Resolution 1986/9, entitled "Use of Scientific and Technological Developments for the Promotion and Protection of Human Rights and Fundamental Freedoms," inviting "The United Nations University, in co-operation with other interested academic and research institutions, to study both the positive and the negative impacts of scientific and technological developments on human rights and fundamental freedoms." The hope was expressed that the United Nations University would inform the Commission on Human Rights of the results of its study of the question.

The United Nations University, in response to this invitation, decided to undertake a study, the object of which was to develop a conceptual framework which would enable the discernment of both the negative and the positive impacts of scientific and technological developments on human rights and fundamental freedoms. The study was to focus on the interaction between socio-cultural, economic, and political factors on the one hand and scientific and technological advances on the other, especially in the developing countries.

Such studies would in turn have two broad aspects. We would need analyses of the ways in which science and technology have advanced the cause of human rights, by bringing within the reach of the majority of humanity many human rights (e.g. in the spheres of health, communication, education) which could in the past be enjoyed only by a select few. Scientific and technological advance has also now made possible the wider enjoyment of these human rights. We need to research ways in which we can extend the reach of these benefits so that they serve much larger segments of the global population than they do at present.

Even more importantly we need studies of the ways in which further advances in science and technology can be harnessed in the service and furtherance of human rights. This is a matter of planning, based on the philosophy that science is humanity's servant and not its master and that the users of science can and should play a role in shaping its directions and using its products.

The attitude of concentration on the adverse impact on human rights of new scientific and technological advances has in it an element of acceptance of the inevitability of the directions of these advances. This attitude implicitly assumes that since these new advances must inevitably occur, we must warn against their dangers, but that we have no control over the directions they take. To the extent that such an assumption is implicit in prior attitudes, it needs to be corrected. We, the users of science and technology, must also contribute an input into the decision-making process which determines what directions they will take.

While it would be unrealistic to deny altogether that science and technology sometimes go forward with a momentum of their own, they are also amenable to direction and guidelines of service to the community, for science and technology are funded externally – in the last analysis by the community – and the community, as funders and supporters of technology, is entitled to a voice in the directions it takes. The need for heavy funding makes it no longer true that the individual scientist or technologist charts out his own course of action as a totally free determiner of the nature and ends of his research.

If this be true we have the opportunity to guide science and technology in the direction of greatest service to humanity. The furtherance of human rights is one such service. Indeed, if human rights represent the quintessence of human achievement in the direction of equality and freedom, there are few higher purposes that science and technology can serve. We have not given adequate thought to this area before, and through this study hope to provide an impetus for further research and more effective measures directed to this end.

Another area of tacit resignation to inevitability is in regard to the tendency of science and technology to serve the affluent sector of the global population. The poor countries are for the most part recipients of technology developed in the affluent world which, as a whole, generates its own technology. As the generator of that technology, the affluent world naturally produces what suits its own requirements or purposes rather than those of the poor world. Indeed, the totality of scientific talent devoted to serving the needs of the impoverished is but a minute fraction – perhaps less than 1 per cent, according to the Brandt Report – of the totality of the scientific enterprise. In spite of the fact that the conditions and needs of the developing world are often far removed from those of the developed world, technology geared to the latter is often the only technology available to the former, however unsuitable it may be.

Science and technology give power to those who wield them, whatever the field involved. Whether it is in the field of communications technology or of bio-medical advances, every new technology gives new power to its controllers. This constitutes a new source of power for the affluent world from which these technologies come and, in the nature of things, such power tends to be wielded in the interests of those who command it.

This tendency is inevitably present and offers the global population another wide area for rethinking old assumptions and attitudes. This study seeks to examine some of these processes by looking at the structure of the scientific enterprise and the points at which more significant inputs can be made by the users of technology. The developing world is the largest eventual consumer of science and technology. In studying ways of deriving the maximum benefit from science and technology, we ought particularly to consider the needs of the developing world and the impediments to technological development.

The problems are numerous. Attitudes have to be altered, priorities determined, new procedures and structures fashioned. Ethical codes for scientists need to be developed, attitudes of resignation and lack of participation in matters technological need to be corrected. Assumptions that Western technology is synonymous with progress need to be revised.

This volume is of course only a preliminary study. It does not aim at providing solutions but rather at setting the background for further studies and helping to identify areas requiring more scholarly effort. Its intention is not to chart out new extensions of human rights but, rather, to find ways and means of implementing existing human rights norms within the changing context of rapid scientific and technological progress. It aims at assisting in the evolution of scientific and technological policies which the developing countries should adopt so that science and technology can be developed in accordance with the needs and requirements of human and social development. It takes its place alongside the research projects launched by the United Nations University in the field of scientific and technological development, such as food and nutrition policy, energy planning, food and energy nexus, biotechnology, and microprocessors. Special United Nations University projects which are more closely related to the present study are those on Technology Transfer, Transformation and Development, Technical Research and Development Systems in Rural Settings, Sharing of Traditional Technology, Self-reliance in Science and Technology for National Development, and Technical Capacity and Prospectives for the Third World.

As this study proceeded it became clear that three categories of human rights related to scientific and technological development should be developed:

1. The right of *protection* against the possible harmful effects of scientific and technological developments.
2. The right of *access* to scientific and technological information that is essential to development and welfare (both on the individual and collective levels).

3. The right of *choice*, or the freedom to assess and choose the preferred path of scientific and technological development.

Of the three categories, *protection* covers the negative aspects, while *access* and *choice* relate to the positive aspects of scientific and technological developments.

It became clear also that science and technology cannot be treated only as products or processes, but should rather be treated as information generated, processed, transmitted and transformed in the process of creating products. This means that we must trace the genesis of such information, its generation, transformation, combination, transfer and application, analysing systematically both the conditions and consequences of interactions between human and social settings and technological information. This would have to be traced through many phases – research and development, commercialization, and technology transfer, to mention only a few. The legal aspects of such processes are often quite remote from conventional concepts of human rights. Research would need to be undertaken to bring the applicable legal norms into closer conformity with developing human rights concepts and to bridge the wide gulf now existing between the two disciplines.

There will be a need for further studies following on from those in this volume. These would examine selected cases connected with conventional and new technologies in relation to human rights, namely: (a) conventional technology: (i) industrial; (ii) agricultural; and (b) new technology: (i) microinformatic; (ii) biotechnology (including bio-medical); and (iii) new materials.

There would also be the need for case-studies related to some of the above technologies, with special reference to:

1. The socio-cultural, economic, political, and ethical aspects of the process of scientific and technological development and the conditions required to guarantee *access* and *choice*.
2. The interface between the process of scientific and technological development and human rights.

Policy recommendations would follow from these analyses of the process of scientific and technological development. The recommendations will need to make an impact in both social and scientific terms. They should also be such that they are accepted as relevant to human rights concerns and capable of being applied accordingly by the legal profession. To ensure this, the project will need to survey the perceptions and opinions of jurists, not only at the level of international experts, but at the level of human rights lawyers, especially in the developing countries.

Such an endeavour cannot of course be undertaken merely on the basis of futuristic studies in one or two disciplines. We need to bring together the wisdom and experience of many disciplines – not merely science, technology, and human rights. We need to survey the broad historical processes involved and the political, sociological, and economic factors at work. We need to harness the insights of philosophy and jurisprudence and to take account of current concerns regarding the environment, the survival of the ecosystem and of the species. Advances in the biological and genetic field raise issues regarding the integrity of the human body and the very nature of humanity.

The studies involved are vast and this volume is only a first step. It seeks to bring together a series of studies from many angles. After this preliminary chapter, we proceed in part 2 to deal with some major global issues which stress the importance of prospective studies, technological self-reliance, and conflicting values and pressures. Such macro-studies provide a setting against which the problems involved can better be analysed. Although some of these studies concentrate, in accordance with our mandate, on the problems of the third world, they contain reflections and conclusions which are of value to all sectors of the world's population.

The volume proceeds in part 3 to examine the international response to the problem in a theoretical as well as a practical sense, in two chapters – one dealing with the normative response and the other with the institutional response.

Part 4 looks at three selected specific issues. These are only samples chosen from a multiplicity of possibilities.

The studies appearing in this volume make it clear that the problem is not one which can be permitted to drift any longer. The progress and course of science and technology are not automatic and inexorable. At every level a number of decision-makers are involved – whether they be multinational corporations, large or small enterprises, politicians, bureaucrats, research foundations, universities, or consumer groups. At the level of choice – whether it is choice of research or choice of product – the developing world must project itself in ever-increasing measure. To do so it needs a complex background of interdisciplinary information and a heightened awareness of the need to interpose its own will rather than let the forces governing science and technology have a free run in its territory.

Scientific and technological choice suitable for a developing country requires much more than a formal structure of local scientific institutions. It needs this kind of interdisciplinary background, and perhaps the present project can help to fill the gap in this area. The important process of technological determinism requires far more information than the scientific community alone can give, however dedicated and well-intentioned they may be.

Science and technology need to be bent in the direction of service; and an important field of general service to humanity is the area of human rights. Seeing that science and technology are among the most potent forces at work today in shaping the way in which we live, it is perhaps opportune that on the fortieth anniversary of the Universal Declaration – the most important landmark in history of the universal acceptance of human rights concepts – we should give thought to the ways in which we can better understand and more thoughtfully direct the forces of science and technology towards the greater protection and fulfilment of the human rights of all citizens of our one world.

We are at one of the great watersheds in history, for whatever course technology may take, we are moving into an era that will be dominated by technology. Whether humanity will live in sunshine or in shadow may well depend on whether it rules or is ruled by technology. Just as technology can light up our future it can condemn us to live in the darkness of its ever-lengthening shadow. It is critical therefore that we make our choices now and that the vast bulk of the world's people realize that it is there to serve rather than be served.

Further delay in the process of analysis initiated in this work can only mean that technology tailored essentially for the developed world will continue to flow and harden in pre-set grooves fashioned for it by the developed world. Every new development presents a new range of choices for further development, but the choice of new development is still dictated by the needs of the developed world. The developing world needs a fuller realization of the range of choice and the extent of scope for independent action available to it. This realization is urgent because technology feeds on itself, and unsuitable technology once introduced tends to be perpetuated. Extensions or variations tend then to be made on the assumption that the prototype provides a fixed and unalterable framework to which the changes must conform. The process once started is difficult to arrest, and can lead to wide divergences between the nature of the technology and the needs of the community.

This study will reveal numerous possibilities for study and action. Only some of them can be chosen, but we hope that the research described will facilitate the difficult process of choice.

Part 2

Global Perspectives

2

Science, Technology, and Human Rights: A Prospective View

AMILCAR O. HERRERA

INTRODUCTION

The present concern for human rights and fundamental freedoms – the main formal manifestation of which is embodied in the Charter of the United Nations Organization – arose basically as a reaction against the widespread violation of human rights in our century. This genesis explains why the studies in this field have been predominantly oriented towards the definition of specific human rights, in order to make it possible to incorporate them into enforceable legal regulations.

As a result of that approach, which may be described as "defensive" or reactive, there is a strong tendency to concentrate action and studies on the possible negative impact of social activities on already defined or accepted human rights, rather than on the new opportunities and options offered by the present process of world transformation.

The above approach is undoubtedly a useful and necessary one, but a study relating human rights to scientific and technological development in the context of one of the deepest crises in human history requires a somewhat wider perspective. A brief look at the past will be of assistance in understanding the present situation in the field of human rights and the type of approach required for the proposed study.

The concern for the fundamental freedoms – all human rights are ultimately dependent on the concept of fundamental freedoms – is one of the constants of history. According to Hegel the history of the world is none other than the progress of the consciousness of freedom. For other historians, freedom is not a product of history; man is born free to work out his own destiny. Whatever our starting-point, however, the political problems posed by man's freedom in society, basically the relationship of the individual to the state and to his fellow men, generate a variety of questions – the relationships between freedom and equality, freedom and justice, freedom and the rights of the state, freedom and law – which have had different answers in different cultures and at different historical moments. Although the philosophical or theological conception of freedom has common roots in all cultures, the way in which that conception is translated into specific institutionalized human rights is, as we know, historically determined, and changes with the evolution of cultures and social systems.

The present conception of human rights and fundamental freedoms was originated by the Enlightenment in the seventeenth and eighteenth centuries. For the Enlightenment, all things in nature are disposed in harmonious order, regulated by a few simple laws, in such a way that everything contributes to the equilibrium of the universe. The same rational order is the basis of the human world and manifests itself through the instincts and tendencies of men. The main obstacle to this linear unending human progress is, for the Enlightenment, ignorance, and the education of all strata of society in the light of reason and science will finally lead to a perfect and happy society. This doctrine underwent a complex evolution during the nineteenth century, but its main premises still linger at the heart of liberal sociology and free-market economics.

Together with liberalism, the most influential version in our time of the vision of history centred on progress is undoubtedly Marxism. In the Marxist conception, the advent of the classless society, through the struggle of the proletariat, would mean the culmination of history, or perhaps more appropriately the beginning of the true history of mankind.

These interpretations of history have a central tenet in common; history is a progressive process governed by internal laws – an immanent natural order, the development of the means of production – whose culmination would be the liberation of man, the creation of a society based on "Rational Freedom," the transition from the "realm of necessity to the realm of freedom." In liberalism and Marxism the promised society is in the future and its attainment will require deep changes in present society, but these optimistic teleological views have a basic faith in mankind, in man's capacity finally to build a free, harmonious society.

In our time, the spiritual climate of the eighteenth and nineteenth centuries has all but vanished. The twentieth century has strongly questioned the central views of the Enlightenment, the idea that progress based on reason – in the restricted sense of the term – and scientific knowledge would lead naturally and unavoidably to a society centred on democracy, justice, and equality. After almost three centuries of material and scientific progress, we face a world situation which is the antithesis of the promised land of the age of Science and Reason.

The great powers have built a destructive nuclear system which, although it can annihilate mankind several times over, keeps growing at a breathtaking pace. In principle – and within its essential irrationality – it could be admitted that these powers can accept their total destruction rather than admit the predominance of the other side; after all, the capacity to decide whether or not to activate the system is in their hands. But the eventual destruction of the population of the third world – about three-quarters of the planet – is a mere by-product of the confrontation between the great powers. They will suffer the effects of a nuclear war without ever having been consulted, without ever having a chance to defend their right to live, or to be spared the consequences of a conflict which is alien to their interests or to their conception of the world.

In absolute terms the gap between rich and poor – measured in terms of the material level of living – has never been so great as it is today between the developed and developing countries. As important or more important than its absolute value is the change in the character of the gap produced in the post-war period

Up to the Second World War the development objectives of developed and developing countries were essentially the same, despite the differences in the starting-points: to increase the welfare of the population through the satisfaction of the fundamental needs, such as food, health, education, housing, etc. After the war, and above all in the last two decades, changes have appeared in the character of the gap which have made the economic indicators increasingly inadequate to describe it. The rich countries confront the problématique of the so-called "post-industrial" era, while the countries of the third world have not yet realized the benefits of the Industrial Revolution, with a considerable part of their population living in conditions of utter poverty. The gap, which was essentially quantitative – leaving aside the cultural differences – is being transformed into a *qualitative* one, with the result that communication between the two blocs into which the world is divided is becoming every day more difficult.

Until the end of the Second World War the most glaring violations of human rights were committed by the developed countries: two world wars, Nazism and Fascism in Europe, colonial domination and violence. During recent decades, there has been a tendency to consider that third-world countries are responsible for a considerable part – or most – of human rights violations, while in most advanced countries those rights are fairly respected.

That vision of the situation has some truth from the point of view of the identification of who is mainly affected by the violations, but not from the point of view of who or what is responsible for them.

The social violence and the repressive regimes of many countries of the third world have their origin, as we all know, in the conditions of utter deprivation in which a great part of the population of those countries lives. Without ignoring the existence of internal elements which help to maintain misery and oppression, we all know that the structural cause of that situation is the asymmetric relationship between the developed and developing countries, created by a long period of political and economic domination. Consequently, the ultimate responsibility for the violation of human rights in the poor countries has its roots, not in specific traits of those countries, but rather on the central characteristics of a world order built primarily by the dominant powers.

The case of the external debt of the third world countries is just an example of that responsibility. Those debts, as is well known, were the result not only of the development needs of the poor countries but also, and in some cases primarily, of a period of rapid expansion of financial capital, that forced the central powers to make heavy investments abroad.

Now, as a consequence of the world recession, the third-world countries confront an external debt whose service imposes a crushing burden on their economies. On the conditions imposed by the creditors – which include heavy interest rates unilaterally determined – the debt can only be serviced at the price of imposing more sacrifices on an already deprived population. Up to now the big powers have shown a total disregard for the terrible social cost of their policies regarding such debts.

The facts that we are referring to do not include anything new, and our purpose is only to emphasize something that we tend, or wish, to forget; that the present social and international world order is, to a great extent, *incompatible with the full exercise of what we consider fundamental freedoms and human rights*.

That incompatibility helps to explain the defensive or reactive approach of the policies on human rights, the efforts to incorporate them into enforceable legal regulations, and the tendency to concentrate studies on the possible negative impact of new social activities, rather than on the new opportunities and options they could offer.

HUMAN RIGHTS AND NEW TECHNOLOGIES: A PROSPECTIVE VIEW

The Specificity of the Present World Crisis

The exploration of the possible risks and of the opportunities and options offered by the new technologies should start, in our view, from two basic premises: the first is that the impact of the new wave of innovations on society can only be properly evaluated in the context of the present world crisis, or, to put it better, of the current process of transformation. The second premise, closely related to the first, is that the character of the social impact is not solely determined by the nature of the technologies but also, and mainly, by the socio–economic strategy adopted to incorporate them.

In relation to the crisis, the fact that the well-known Kondratiev-Schumpeter theory – which associates economic crises with waves of innovations – refers to cycles and to a recurrent phenomenon stimulates a dangerous tendency to predict the evolution of the present crisis on the basis of past experience, particularly the crisis that culminated in the 1930s. This approach does not take sufficiently into account the fact that the process of change that each crisis represents has a specificity which cannot be understood simply in terms of incremental changes in a constant set of more or less quantifiable variables. There are elements of discontinuity which, although difficult to quantify, play an essential role in the evolution of the crisis.

In our opinion, the main elements that differentiate the present crisis from the previous ones are the following.

The Emergence of the Third World

In the 1930s the world was broadly divided into the countries we now call developed – basically Europe, the USA, Canada and Japan – and a vast conglomerate of countries, most of them colonies, with little participation in the world structure of power, and whose economic role was basically to export raw materials, and to import manufactures from the industrial powers.

The third world – a result of the post-war reorganization – is now an active protagonist on the international scene and cannot be disregarded by the major powers. Some of the most important political events of this century, due to their short- or long-term repercussions – such as the Chinese and Cuban revolutions, and the Viet Nam war – have had as protagonists countries of the third world. Central America and the Middle East are only two examples of regions of the third world whose problems directly or indirectly affect the world power structure.

From the point of view of the world economy as well, the third world is a presence that cannot be ignored. As is well known, the enormous external debt of the developing countries is one of the determining factors of the future evolution of the international financial system.

The Emergence of the Socialist World

In the interwar period the only socialist country was the Soviet Union, relatively isolated and with little direct influence in the world power and economic structure. Now the post-war expansion of the socialist bloc in Europe and the presence of China – besides smaller countries such as Cuba, Ethiopia, and Viet Nam – have converted the socialist world into a critical element in the future evolution of the international system.

The recent changes in the Soviet Union and previous events in other socialist countries are clear manifestations of a process of internal evolution which is not less important because it is only sporadically visible. Besides, the growing trade relations not only with Western Europe, but also with other regions or countries, indicate that the influence of the socialist world increasingly transcends the purely political and military spheres.

Questioning of the Present Values of Western Society

Until no more than two or three decades ago there was the general feeling that the process of world unification indicated by the expansion of the Western powers, and enormously accelerated after the war, meant essentially "world Westernization." The colonization of most of the world and, more recently, the one-way transference of technology, and the diffusion of Western-style industrialization with its implicit cultural values, seemed to condemn to almost complete obliteration the achievements of other cultures.

This process is starting to change, firstly because the Western world has begun to have serious doubts about the soundness and rationality of its own conception of progress and development and, in its search for alternatives, is becoming aware that other cultures can perhaps make decisive contributions to a more integrative, less reductionist vision of the world; and, secondly, because the other cultures have started to assert their own identity and to reject a supposedly universal concept of development which does not take into account their own cultural specificity.

The New Wave of Innovations

The new innovations belong to several technological fields – micro-electronics, biotechnology, materials, energy – but what gives them the character of a "wave" is the fact that they tend to be mutually articulated into a "cluster" which defines a new global technological paradigm. The central element of the cluster, the one which determines the character of the new paradigm, is micro-electronics.

The dominant characteristic of the new wave is that its impact seems to be more important to the organization of production, the labour process and the social division of labour, than to the general profile of the productive system. The Industrial Revolution, with its first great modern wave of technological innovation and the emergence of the proletariat, consolidated the capitalist economy and changed Western society. The subsequent technological waves changed the whole profile of the productive system, but did not alter signficantly the structure of capitalist society. This new wave will affect the very basis of industrial society, as can be seen by considering briefly the process of automation and robotization.

In all modern societies access to goods and services is conditioned essentially by wages in the widest sense – the remuneration of labour in any of its forms. In the future this central role of wages will decrease, firstly because one of the consequences of automation – the elimination of most jobs that do not require "non-programmable" skills or creativity – will obliterate most significant forms of hierarchy in the labour process; secondly, because direct participation in the productive system will become a diminishing fraction of total human activity, and so its importance as a determinant in the distribution of goods and services will be greatly reduced. The transition to the new "mode of production" will undoubtedly take some time to be completed – of the order of one or two generations – but its first effects are already with us.

The institutional response that the advanced societies are making to the growing problem of unemployment generated by the new technologies is based primarily on the payment of a "salary" to the unemployed through social security services. The non-institutional response is a rapid growth of the service sector, and of the so-called informal sector of the economy; in both cases the apparent solution only conceals the real problem. Most of the workers "transferred" from industry to the service sector – as is well known in the United States, for instance – perform menial tasks in restaurants, coffee shops, and the like, with salaries that are only a fraction of those they earned in their previous jobs. As for the informal sector of the economy, it includes a high proportion of "independent" workers who participate marginally in the productive system, deprived of any kind of social protection, whether from government or trade unions.

In the case of Western Europe, it can be estimated that at least 25 per cent of the economically active population – unemployed or barely surviving in the informal sector – is socially marginalized. So the phenomenon of social marginality, a consequence of structural unemployment – which was a characteristic of the underdeveloped countries – is now appearing in the developed world. The main difference is that the advanced countries can afford a much higher degree of economic protection to the unemployed, but the phenomenon of social marginalization is essentially the same.

There are other areas of impact of the new technologies which have important consequences for society. Nevertheless, in this brief presentation we have decided to concentrate mainly on the automation and robotization process because, as we have already said, that is the most important social impact and the one that best illustrates the impact of the new technological wave on the field of human rights.

The Environmental Limits

The consciousness that natural resources and the environment constitute absolute limits to economic growth only appeared in the 1960s. We know now that material consumption cannot grow indefinitely without taking into consideration its effects on the equilibrium of the biosphere.

That awareness, however, is not reflected at the high levels of social decision-making, where a deep ambiguity prevails in relation to environmental policies. Never in history has mankind had the capacity to forecast the results of its actions as it has today. The enormous amount of information accumulated at world level by national and international organizations and the modern means to process it make it possible to have, if not an accurate long-term picture, at least the general trend of some of the variables which condition our future. Yet, there has probably never been a greater inconsistency between a predicted future and the measures taken to cope rationally with it. We have become aware that the resources of the earth are finite, but we still consider – above all in the capitalist world – indiscriminate economic growth to be the universal panacea for all our social and economic ills.

The Destructive Nuclear System

All elements of the crisis mentioned above imply the possibility of conflicts. The form and extent of those conflicts is conditioned by the fact that we have now a nuclear destructive capacity ready to be fired equivalent to about a million Hiroshimas.

The crisis of the 1930s did not end due to the application of Keynesian economic measures; it ended as a consequence of the Second World War. A global war could also put an end to the present crisis, but through the destruction of mankind and of most of the biosphere in which we live. Whether the physical annihilation of our race would be complete or not does not matter very much. There may be survivors, but all we associate today with humanity and civilization would have been totally obliterated.

Besides the continuously increasing danger of collective suicide, the cost of the arms race is one of the obstacles to the solution of the problems associated with poverty that affect a great part of the world. In 1985 the global military expenditures – 940 billion dollars – exceeded the total income of the poorest half of humanity.

The first five elements we have briefly examined show, if we consider this crisis in the context of the theory of the cyclical long waves of the capitalist economy, that the present crisis has a character which makes it very different from previous crises since the beginning of the Industrial Revolution. The last two elements – the physical outer limits, and the destructive nuclear system – show that this crisis has no precedents. For the first time in history, humanity can be destroyed by its own actions.

It can be said that our civilization has become dysfunctional in the sense that it is no longer able to make adequate responses to the problems generated by its own evolution. We are facing a global crisis whose trajectory, due to its enormous and growing complexity and its lack of precedents, cannot be simply deduced from the past. Viewed as a continuous process, it has some of the characteristics of a discontinuity in the evolution of human society.

We must conclude that we are in an extremely complex situation. We are confronting a future whose evolution is very difficult to forecast and, at the same time, we need some guidelines for action *in the long term*. The only solution to our present predicament is to formulate and implement alternative development strategies based on objectives more in accordance with the aspirations of the majority of the population, with the possibilities and constraints posed by the advance of our scientific and technological knowledge, and with our understanding of the physical universe in which we live.

We do not pretend, of course, that to stress the need for long-term prospective studies constitutes an original proposal; the widespread perception of the need for this type of work is demonstrated by the well-known long-term global forecasting studies initiated in the 1960s. There is, however, much less agreement about what type of prospective approach is really relevant for the present world situation and this is a central point to be discussed in connection with research in the prospective dimension.

TENDENTIAL OR NORMATIVE FORECASTING: A CRUCIAL OPTION

The selection of the type of approach to be used in the proposed prospective studies is not simply a technical or academic problem: it is also, and mainly, an option between conflicting conceptions of how to confront the present process of world transformation. The best way, in our view, of understanding the significance of the choice to be made is to analyse briefly the criteria used – explicitly or implicitly – in selecting the approaches to be applied in the prospective studies. The fact that the best known of the long-term forecastings were embodied in simulation models, and that the prospective studies we are referring to have a much wider scope, is irrelevant to the analysis: the basic philosophical options in both cases are the same.

The series of long-term global forecastings started in the 1960s had two contrasting views of the future. H. Kahn's *The Year 2000 – A Framework for Speculation on the Next Thirty Years*, published in 1967, presented an optimistic view of the future, without important discontinuities or qualitative changes. In Kahn's view, "despite much current anxiety about thermonuclear war we are entering a period of general political and economic stability, at least so far as the frontiers and economies of the old nations are concerned."

The other current of thought in forecasting sees, instead of a future of unending progress, humanity rushing towards an almost unavoidable catastrophe. In this view neo-Malthusian ideas are combined with the modern concept of "outer limits." This world-view has its best-known advocates in Anne and Paul Ehrlich (*Population Resources and Environment*, 1970, and *The Population Bomb*, 1971), J. Forrester (*World Dynamics*, 1971) and Dennis Meadows and co-workers (*The Limits to Growth*, 1972).

For those forecasting the growth of population and consumption with the ensuing pressure on natural resources and the physical environment, humanity is heading for a disaster that will result in a sudden decline in population and a miserable level of living for the survivors. Largely as a reaction to the "models of doom," several models appeared in the early 1970s, the best known being the Fundación Bariloche World Model and the Mesarovic–Pestel Model.

The second generation of global long-term forecasting appeared at the end of the 1970s or the beginning of the 1980s. The most important among them are *Interfutures* (OECD), the *Presidential Report on the Year 2000*, and the *Brandt Report*. All these studies start from the present situation and try to describe the future, or a range of possible futures, on the basis of the assumed predominance of present observable trends.

From the point of view of their basic approach, these global forecasts have been divided into two groups: tendential (or positive) and normative. The former describes a possible future assuming the persistence of the main tendencies observed at the time. The normative approach proposes a possible and desirable future, and attempts to identify the actions needed to move from the present towards that

future. With the exception of the Bariloche study, all these explorations of the
future would be tendential.

We can now make a comparison – from the point of view of their objectives
and methodology – between what we can call the studies of the North and the
studies of the South. We obviously cannot attempt a detailed analysis in this brief
article; our purpose is only to show the implications of the prospective approach
adopted.

From the point of view of the type of forecasting selected, it is generally
accepted that there is a clear difference between the two groups of works: those
from the North would be essentially tendential, while those from the South
would be normative. In our view, however, despite their apparent methodological differences, both groups of studies use the same basic approach.

In relation to the third world, the findings of the forecastings made in the
North are basically similar. At the beginning of the next century the gap between
the rich and the poor countries will be greater than now, or will diminish marginally in the more advanced developing countries. In absolute terms, the situation
in the poorest part of the third world will probably worsen.

The main implicit premises of the studies of the North are essentially two: the
first is that there will not be essential changes in the present social and international order, although there will be some adjustments and possible changes in the
pattern of distribution of power among the advanced capitalist countries; the
second premise is that the third-world countries will not produce actions that
can alter significantly their present situation in the world economic and political
power structure. In other words, the future of the third world is a dependent
variable of what will happen in the advanced countries.

In the studies of the South the basic premises are different: they start from the
assumption that the present crisis far transcends the economic and technological
dimensions and question the very basis of the present social and world order and
its underlying values. As in all periods of transformation, a wide array of new
options is opened up, and these offer the third-world countries the opportunity to
participate actively in the construction of a new and more equitable world order.

It has been argued that an essential difference between the two groups of studies
is that the ones from the North take as a basis observable present trends, and are
therefore "objective," i.e. do not introduce scenarios based on subjective or
value-based judgements, as would be the case with the South studies. To what
extent is this argument valid?

In the studies of the North the privileged variables, the ones that determine the
state of the system, are mainly economic and technological and they cannot per se
introduce radical changes, or discontinuities, in the global evolution of social and
international systems. Besides, and most important, these variables are to a great
extent controlled by the advanced countries, and are amenable to quantitative
treatment. Hence, through adequate information on their values and tendencies,
the North can expect to maintain a reasonable control over them even in situations of rapid change.

To incorporate the possibility of transformations that can alter the "tendential" future, it is necessary to consider also the social actors, the ultimate agents of change involved, and this is what the South prospective studies do. In the Bariloche Model, the proposed society is supposed to represent the will and aspirations of the majority of the population. The Technological Prospective for Latin America Project, now in its final stage, is based on socio-economic scenarios and selects desirable options among the range of possible futures determined by the interplay of the social actors – national as well as international – involved.[1]

It is clear, therefore, that a crucial difference between the two groups of studies lies in the choice of trends – represented by variables – and that selection has to be explained by factors other than whether or not these are observable or "objective."

In our view the selection of variables is mainly determined by the fact that the countries of the North have a privileged position in the world power structure, and so it is only natural for them to avoid or underestimate variables – over which they have little control – that can alter an already unstable situation. In a position of privilege, any change is potentially dangerous and to ignore or minimize it has been a recurrent attitude all through history.

In conclusion, the basic approaches followed in the studies of the North and the South are not so different as they seem to be. Forecasting, starting with the specific attitudes of its authors, selects, through the choice of variables, one or a set of options among a whole range of possible futures. To assume that present trends will continue into the future without significant changes is, in present circumstances, at least as "normative" or "subjective" as to assume that these trends are not viable in the long-term perspective. In both cases there is a choice of a future: the basic difference between them is that in the case of the North the selected future is the continuation of the present situation with the minimum possible degree of change, while in the case of the South it is assumed that continuation of present trends is neither viable nor desirable, and a viable scenario is selected from among a multiplicity of possible options on the basis of explicit value judgments.

These are not "objective" visions of the future because there is no predetermined future; there are only options. Forecasting is as much a tool for shaping the future as it is an instrument for exploring it. This means that forecasting is not simply a theoretical exercise, but is always performed – implicitly or explicitly – as a guide for action. The selection of variables implies the selection of long-term objectives, and it is only natural for the forecasters to highlight the variables which, in their view, hold the key to the attainment of the proposed goals. Thus the divergence between the prospective studies of the North and the South reflects the conflicting relations between the two blocs into which the world is divided, rather than methodological differences in the strict technical sense.

We do not want to leave the impression, however, that we believe that it is only in the third world that the viability and desirability of the present world order is being questioned. In previous analyses there have been unavoidable simplifications, the most important being the apparent clear-cut division between the North and the South perceptions of the world. The attitude towards the future embodied in the North studies is far from universally accepted in the developed countries. There is an important part of the population – particularly significant in the younger generation and among intellectuals – that strongly questions the vision of the world implicit in these prospective studies. At a more general level, an attitude of confrontation is one of the basic elements of the ecological, peace and feminist movements and, as we also know, a most important part of the literature on alternative futures is being produced in the developed countries. We used those studies as representative of the North, at the risk of oversimplification, because they reflect the position of a great part of the upper levels of political decision-makers, those that have a central responsibility for the shaping and maintenance of the present world order.

In conclusion, what we need as a guide for actions that could give us hope for avoiding the dire consequences implicit in present world trends are long-term normative prospective studies in the strict sense – in other words, studies that focus, among the possible options, on future scenarios that are viable and desirable. It is in this frame of reference that the impact of new technologies on human rights and fundamental freedoms can be properly explored.

NORMATIVE PROSPECTIVE STUDIES AND THE TECHNOLOGICAL DIMENSION

We can consider now the role of the technological dimension in the prospective studies we are referring to. Our discussion will be based on the methodology used in the Technological Prospective for Latin America project.[2]

The central feature of the TPLA in relation to the subject under discussion is that the frame of reference for the formulation of scientific and technological strategy is the R&D demand of the desired society. This means that R&D policy should not be determined by specific problems, or areas of problems posed by the technologies per se, but rather by the socio-economic, political and cultural goals proposed by the chosen society.

This approach implies the following sequence of steps:
1. Definition of the social goals to be reached.
2. Identification of the obstacles to the attainment of those goals.
3. Formulation of a socio-economic and political strategy for overcoming those obstacles.
4. Determination of the scientific and technological demands of the strategy.

It seems an almost hopeless task to reach a general agreement on what a "viable and desirable society" means, but this is a necessary pre-condition for the attainment of a just and stable world order. In the TPLA project it was decided to define the desirable society on the basis of a few normative characteristics defined as "invariants," in the sense that if any one of them is not present, the society is not desirable. This allows for a multiplicity of societies which are desirable, within a wide spectrum of cultural and organizational differences.

The invariant characteristics adopted for the desirable society are the following:
– Essentially egalitarian in the access to goods and services.
– Participative: all members have the right to participate in the social decisions at all levels.
– Autonomous (not autarchic).
– Intrinsically compatible with its physical environment.

These characteristics may seem too general, but they are enough to define a basic type of society and, more important, they are shared by the majority of mankind. They are what can be called first-order long-term goals. They constitute the frame of reference for the formulation of the short- and medium-term objectives. These objectives could vary greatly, depending on national or regional conditions and on the selected strategy, but they should fulfil the objective of contributing – or at least not hindering – the final attainment of the first-order goals.

We cannot attempt a detailed discussion on how to deduce the social technological demand from socio-economic strategy. We will only make some brief general comments on points that could be particularly relevant to the subject under discussion.

The traditional way of formulating an R&D strategy – to identify the demand for goods and services over the productive system and from that to deduce the demand over the R&D system – should be considerably changed.

That approach, which is essentially sectoral, does not allow the adequate identification and incorporation in the R&D demand of the *qualitative* elements – such as participation, relations with the environment, decentralization of decision-making and production, cultural traits, autonomy, etc. – which are so important in characterizing a society.

An R&D strategy is not simply the addition of the sectoral demands, taking them in isolation; it is necessary to articulate them and to establish priorities. On the other hand, the interaction between the sectors and, consequently, the factors which determine priorities vary with time, above all in a process of transformation as deep as the present one. This is particularly evident in the qualitative elements already mentioned. It is obvious in consequence that essential components of the R&D demand cannot be identified through the sectoral approach because they are conditioned by the global evolution of the society. In other words, the formulation of the R&D strategy requires a reasonably clear conception of the possible global evolution of the society in the period under consideration.

That approach means that the present dichotomy in practice between socio-economic and R&D planning should be overcome: the scientific and technological dimensions should be explicit variables incorporated in the whole process of socio-economic planning.

The integration of socio-economic and scientific and technological planning in the context of a changing society cannot be effectively attained unless there is close interaction between social scientists, technologists, and natural scientists. So the implementation of real interdisciplinary research – instead of the loose addition of knowledge from different disciplines that we now call interdisciplinary research – is one of the most difficult challenges confronted by the social sciences today. We cannot discuss this complex subject further; we only want to stress that the basic pre-condition for really relevant co-ordination is to start by *posing* the problems in an interdisciplinary context, instead of the present common practice of defining a social problem in terms of a single component – economics, for instance – and asking for the support of other disciplinary fields afterwards.

The approach to the study of the impact of the new technologies, so briefly outlined, has in our view the following advantages which are particularly relevant for third-world countries:

1. It eliminates the danger of falling into what can be called the "defensive" approach, which consists of the identification of possible negative impacts of the new technologies – for instance, unemployment – and the concentration of effort on the avoidance of those supposedly negative effects. The identification of the R&D demands in the context of socio-economic strategy allows an objective and unprejudiced evaluation of the opportunities and risks posed by the new technologies.

2. It allows an easy and "natural" articulation of the new technologies with the current technologies, including the traditional ones. One of the central elements to be taken into account to identify the R&D demand is the present technological context.

3. It should be emphasized finally that, although the R&D demand is identified at the formulation of the R&D strategy, the potential of the new technologies is taken into consideration in the establishment of the socio-economic objectives. There are socio-economic objectives that were not possible in the past, but are possible now owing to the new technologies.

The New Technologies and Human Rights and Fundamental Freedoms

On the basis of the methodology outlined above, which attempts to define the socio-economic and political strategy required to achieve a society compatible with the full exercise of the fundamental freedoms, we can now attempt to explore the relationship between new technologies and human rights. The full understanding of that relationship will take an effort which goes far beyond the possibilities of a single project. There are some preliminary results, however, that could help to identify lines of research which deserve further exploration.

The first one, as we have already pointed out, is that the most important social impact of the new technologies is the impact of micro-electronics – through automation and robotization – on the organization of production, the labour process, and the social division of labour. The problem is not whether or not traditional forms of work and employment will be abolished – that change is inherent in the transformations induced by the new technologies – but rather the way in which they will be abolished.

The growing recognition that the character of present unemployment confronts the advanced countries with a problem that cannot be solved without a complete questioning of the relationships between technology, employment, and work is starting to generate a debate on the subject.

In a recent paper, W. Zegweld makes a proposal that, although still very general, is an advance in the right direction.

The problem is to organize the breaking down of barriers between traditional wage-earning employment and work in the widest sense of the term. Such work can provide income but also offers a social role, contact with others, and opportunity for creation or enterprise. It must not be proposed in a single, rigid setting identical for all but must be flexible enough to meet the wide variety of demands and respond to freely expressed choices. Instead of offering everyone a problematical full-time job, the aim is to allow everyone to find and choose a job in which working hours, level of pay and social security coverage are no longer pre-determined and closely linked but can be adapted, above an indispensable minimum, to wishes of the individual . . .[3]

In A. Gorz's opinion, the characteristics of socially necessary work do not allow for any creativity or personal development of the worker. The only solution is the reduction of working time through the new technologies and the distribution of the remaining socially necessary work among the whole population. In his view, "the choice is between the liberatory and socially controlled abolition of work or its oppressive and anti-social abolition."[4]

That is the challenge confronted now by society, particularly in the developed countries, and it is still too early to know with certainty how the process of change will evolve. We believe that the first option will finally prevail: first, because the oppressive imposition of new forms of work would be extremely difficult in a society where social control is largely based on the discipline imposed by the traditional relationship between work and employment – precisely the relationship that the new technologies can radically change; and second, because it allows satisfaction of one of the oldest aspirations of our species: the liberation of human beings from routine work that does not require creative capacity.

The solution to the problem of employment applied now in the developed countries – the payment of a minimum for subsistence to the unemployed – creates a category, the *unemployed*, which represents, in practice, a form of social marginalization. In A. Gorz's words:

Whatever the amount of the minimum guarantee, its fundamental vice remains: it leads to a gash in society, to a dualist stratification that can amount to a South Africanization of the social relationships. The minimum guarantee is really the salary of the social marginalization and exclusion. . . The minimum guarantee is a way of accepting that gash, and of consolidating and making it more tolerable.[5]

In the social project of the TPLA the employment policy is based on the principle that every person has not only the duty but, above all, *has the right to a useful task in society*.

In the first stages of development of the proposed society, all the members of the workforce without a place in the productive system will receive a state subsidy – as happens now in the industrialized countries – that can ensure them adequate access to all basic goods and services. The difference is that they will have to fulfil some socially useful task.

That conception means that it will be necessary to create new socially productive activities outside the traditional forms of employment. This policy will be greatly facilitated by the fact that the process of transformation activated by the socio-economic strategy will certainly generate, or stimulate, social activities which do not have a clearly defined role in the present employment structure, such as informal education, community organization, preservation of the environment, etc.

In more advanced stages of the process of change the objective will be the uniform distribution of the diminishing socially necessary work among the whole population. This will require a complete redefinition of the relationship, work–employment–technology, and of the social role of salary. This policy, made possible by the advances of technology, would mean the end of one of the greatest evils in the history of our species: the social division of labour between intellectual and manual, or, perhaps more appropriately, routine work.

The low relative efficiency of the technologies used in the past meant that most of the population had to devote its effort to the physical tasks connected with the primary needs of material life, so that through the physical work of the majority a small minority could be liberated to devote itself to the social functions which demand a higher input of knowledge and creativity. The alternative solution, the distribution of the lower tasks among all members of the community in order to give everybody the opportunity to participate in the higher functions, seemed not socially feasible because a homogeneous distribution of physical or routine labour would have allowed each individual only a negligible marginal time to be devoted to other activities. As the higher functions in this type of society require a long and intensive training, that amount of individual free time would not be enough to provide for it.

With the advances of the technologies of production after the Industrial Revolution, it became increasingly difficult to justify the injustice represented by the character of the prevailing division of labour. With the coercion made possible through the appropriation of the social surplus, it became necessary to invent a moral justification for that glaring inequality. In the Western world it took the form of a glorification of brute physical work, which has no precedent in history: manual work was exalted as the source of all virtues, as the very foundation of all that man has accomplished. Not surprisingly, the overwhelmingly dominant voices in this universal chorus were those of people who had never had the need to perform any significant amount of routine physical work to earn a living.

Now, as the culmination of a process that started when man invented the first tool to facilitate his interaction with the physical world, it finally becomes possible to formulate a human right, complementary to the one stated above: *the right of every person to have access to intellectually creative work.*

The relationship of man with work will be, in our view, a focus, perhaps the central one, in the discussions in the field of human rights and fundamental freedoms. Up to now, and specially since the Industrial Revolution, for most persons work has been almost synonymous with employment – i.e. work of a standardized, repetitive nature. Employment gave an income to provide for basic human needs. Now we have the possibility of recovering the original meaning of work: the liberation of the creative capacity of man.

Another human right, the context of which has entirely changed with the new technologies, is the *right of participation in all social decisions.* There is a relatively wide consensus – at least at the level of declarations – that in a really democratic society all persons should have the right to participate in social decisions in a more effective and direct way than the periodic election of governments. There is less consensus, however, in relation to the possible mechanisms of participation. In general, there is a tendency to subordinate this basic human right to the attainment of certain objectives which are considered pre-conditions for effective participation: higher levels of education and social awakening, creation of adequate means of access to information, etc.

The implicit philosophy of the TPLA is that participation is not only a means for more efficient social organization, but an end in itself. In the process of social change persons are not liberated when certain specific goals are attained – as for instance, in poor countries, when basic needs are satisfied – but rather when they feel that they are protagonists of the process, and not simply passive beneficiaries or victims.

Starting from the principle that full participation is the result of a process, and that the only way to learn participation is by participating, the process should start at the very beginning of the period of transition, through participation at the levels of the community and workplace. This approach has two main advantages:
1. At the community and place of work levels, all persons have, or can obtain easily, the information required to take decisions on matters that concern them directly.
2. The decisions taken at those levels affect the participants directly; the well-known mechanism of trial and error is the best school for learning a conscious and responsible participation.

It is very difficult to foresee the character of the mechanisms that will emerge to allow participation at higher levels of decision-making. We want, nevertheless, to emphasize the following points:
1. The decisions taken at the community and place of work levels necessarily affect the upper levels.
2. The type of decisions taken at the first level reflects clearly the general lines of thinking and opinions of the population.

3. Taking into consideration the previous points, it is difficult to imagine that upper levels of decision-making would try to impose policies in open contradiction to the tendencies expressed at the first levels. That would generate conflicts – at the operational and implementation levels – which would jeopardize the viability of those policies.

Summing up the above, it could be said that the pre-condition for real participation is that the social actors who conduct the process of change should regard participation not only as a characteristic of the new society, but also as a fundamental instrument with which to build it.

Participation is not possible, above all in societies of the size and complexity of the modern nation-states, without adequate information. The advances in information technology mean, for the first time in history, that the information necessary to make social and economic decisions can be made available to the whole population. One of the goals of the R&D strategy should be to study what is the most efficient way – not only technically but also from the point of view of how participation is socially organized – of making the required information available to the participants.

Participation, besides being a fundamental human right, is essential for the preservation of other rights and freedoms. The new technologies relating to information can make an enormous contribution to the construction of a better and more democratic society, but they can also be used against the exercise of fundamental freedoms by helping to concentrate information – and consequently power – in the hands of the state or of dominant social groups, instead of making it available to the whole population.

Only through the effective participation of the population in all important social decisions can that menace to the fundamental freedoms be eliminated. History shows that the defensive *ex post facto* approach, of defining specific freedoms that could be affected by the new technologies, is not effective enough. Only a society intrinsically compatible with the fundamental freedoms – and this means a participatory society – can ensure a socially creative use of information.

Another field of concern regarding new technologies and human rights is the dependence of the third world on the technological knowledge produced in the developed countries. We cannot analyse this complex subject in detail here but, in our view, any constructive discussion of the problem should take into account the following points.

A central premise of the TPLA approach is that the main problem for third-world countries is not so much to close the technological gap in absolute or abstract terms, but rather to "close" it in the context of the socio-economic and institutional adaptations required for the creative incorporation of the new technological paradigm. In other words, the way the technological gap should be closed is a dependent variable of the socio-economic strategy.

The above leads to the concept of what scientific and technological autonomy really means.

The technological dependence – in greater or lesser degree – is unavoidable, even in countries with R&D systems more developed than those in the third world. This makes the concept of "technological space" in the TPLA project particularly important.

In this approach, endogenous generation of technology refers basically to the process through which the characteristics that a given technological solution should have are determined. The information thus produced – social, economic, cultural, environmental – defines technological space.

All possible solutions that fit the technological space should be considered. In other words, the *endogenous process is the process of definition of the character of the required solution*, and not necessarily the technology itself, which can be imported provided it fits the technological space. In this way the transfer of technology becomes an integral part of the process of generation of technology.

In the determination of technological space, participation has a decisive role: the appropriate technological solutions required for the building of a new society can only be generated through an active interaction between the R&D systems and the social demands expressed through the mechanisms for participation referred to above.

We want to emphasize finally that although we have presented the above approach or methodology to explore the impact of the new technologies on human rights as a view from the third world, we believe that the approach has a validity which far transcends the specific circumstances of the developing countries. The starting-points are different, but the challenge presented by the global process of transformation seen in the crisis is basically the same for the three blocs into which the world is presently divided. Each of the blocs will follow its own trajectory in the period of transition to a new society, but unless these trajectories are convergent, there is little hope of finally building a world order compatible with the full exercise of human rights and fundamental freedoms.

NOTES

1. There are other important regional studies carried out or being carried out in the third world, most of them sponsored by thc UNU. We do not include them in our treatment because their time horizon is shorter and most of them do not include technology as an explicit variable.
2. The research organizations involved in the project are: Centro de Estudios para el Desarrollo, Venezuela; Núcleo de Política Científica e Technologica, Universidade Estadual de Campinas, Brazil; Fundación Bariloche, Argentina; Universidad Nacional Autónoma de México, Mexico. The work is conducted by a Steering Committee composed of H. Vessuri, G. Gallopin, R. Dagnino, A. Furtado, L. Corona and A. Herrera (co-ordinator of the project).
3. W. Zegweld, "Technology, Employment, and Work," paper presented to the International Symposium on Perspectives of Science and Technology Policy, Guanajuato, Mexico, February 1984.

4. A. Gorz, *Adieux au proletariat. Au delà du socialisme* (Editions Galilée, Paris, 1980).
5. A. Gorz, "Qui ne travaille pas mangera quand même," *Futuribles*, July–August 1986.

BIBLIOGRAPHY

Barakova, N., ed. *Unattractive Work*. Georgi Dimitrov Research Institute for Trade Union Studies. Sofia, 1985.
Gorz, A. *Adieux au proletariat. Au delà du socialisme*. Editions Galilée, Paris, 1980.
Rothwell, R., and W. Zegveld. *Technical Change and Employment*. Frances Pinter, London, 1979.
Salomon, J.J., and G. Schméder, eds. *Les enjeux du changement technologique*. Ed. Economica, Paris, 1986.

3

Technological Self-reliance and Cultural Freedom

SANEH CHAMARIK

INTRODUCTION

The main concern of this paper is with the infrastructure and social role of science and technology, with a focus on developing countries in their current efforts toward modernization and industrialization. For all the difference in emphasis and despite the distinctive traits attached to each, science and technology are closely connected on both methodological and epistemological grounds. Together they are related to social and cultural problems.[1] The nature of this interrelationship and its social implications should become increasingly clear in the course of further discussion. For analytical purposes, the two are to be treated as a single and integrated whole. Their dual functions must be examined and mutually assessed.

First, with respect to the physical world, advancement in science and technology can help bring about development in terms of increasing productive capability and greater freedom vis-à-vis the constraints of nature. Secondly, such advancement is also instrumental in producing societal change and transformation, with significant impacts on problems of human and social relations. Hence the specific human and social dimensions of science and technology need to be objectively perceived, quite apart from their technical and seemingly universal character.

It is this specific social context which, by and large, determines the course and pattern of technological development as well as its consequences. Thus the impact of science and technology has to be evaluated on account of both its cause and effect. This is all the more pertinent to the developing countries as latecomers in the field. Most, if not all of them, are somehow bent on following in the footsteps of the West in advancing from the agricultural phase into the industrial and post-industrial phases. The objective and model of development seems clear-cut, at least in the eyes of the third world's modernizing élite; that is, to accelerate economic growth through industrialization. This is their foremost priority, and it is to be achieved by riding on the waves of technological change which the West has already survived and established within its socio-cultural context.

In a significant historical sense, this development trend is a reflection of the enigmatic impact of Western development.[2] This fact adds an international dimension to the problem of the relationship between exogenous and endogenous sources of technological capability and creativity. More often than not, the virtues of modern science and technology are simply taken for granted. They are looked upon as something of absolute value and have thus become, willy-nilly, an end in themselves, politically and ethically neutral and free from any damaging influences. This is the crux of the whole problem. It is by no means a mere question of the use or misuse of science and technology from a purely technical standpoint, but involves the whole spectrum of socio-cultural factors underlying technological growth and development.

In this continuing "dialectic of specificity and universality,"[3] to use a phrase of the noted physicist R.S. Cohen, scientific technology both offers opportunities to some and is fraught with dangers for others. This has been historically demonstrated for Western societies, and is about to take place in societies tied to the same growth model. In view of the human and social costs involved, one can no longer remain complacent about the adequacy of the overall objective of economic growth, as has hitherto been the case. This would represent only a partial appreciation of reality, one that could turn the very virtues of science and technology into a weapon against humanity.

Indeed, to guard against the adverse impact of science and technology, there is an urgent need to set the whole problématique in proper historical perspective and to deal with the specific problem of human and social relations accordingly. As R.S. Cohen again reminds us:

In the attempt to understand the social impact of scientific technology, we must proceed simultaneously in two ways: first in a far less sweeping and generalizing manner (is technology good or is technology evil?), and second, in a far more self-critical and sceptical dialectical analysis (science gives life and death). We also should recognize the historical character of our attitude towards the social and human impact of science and technology within our own century; attitudes towards technology will differ depending on whose technology it is, or which specific technological advance we evaluate, or which portion of humankind is speaking or is represented, which class, which race, which tribe, which generation, which sex, at which cultural place the evaluator stands . . . [4]

Thus, along with technological growth and development, there also emerges the problématique of rights and obligations. It is in the light of its social and historical paradigm that the status of modern science and technology needs first to be re-examined and evaluated. In the process, the concept of human rights itself, as defined under the philosophy of liberalism, needs to be clarified in its historical context. From such analyses, objective principles could hopefully be drawn, especially for developing countries, which are at the receiving end of scientific technology. The entire issue is not confined to any particular countries or groups of countries, but is fundamentally global in character. It thus involves all countries concerned in the process of giving or receiving technology, no matter at what stage of scientific and technological development they may be.

STRUCTURAL NATURE OF MODERN SCIENCE AND TECHNOLOGY

For the purpose of scientific advancement, there is an obvious rationale, and indeed a great advantage, to learning or even borrowing from the Western precursor. It certainly should not be a question of whether or not Western science and technology ought to be made use of, but of how, on what conditions, and for what objective. The answer to all these primary questions involves consideration of the moral and spiritual values that are involved in the concept of progress. If scientific technology is to be used to aid progress, and progress involves moral and spiritual values, the linkage between technology and moral and spiritual values is dear. Hence the functional relationship between technology and human rights needs to be recognized.

While technological advance serves to liberate humankind from the forces and constraints of nature, it is precisely the same scientific knowledge and technical skills that can bring about domination of man over man. Which result ensues depends on the status given to science and technology and its structural relationship within a given society. For example, they could be preserved as a privilege of the few, or widely shared among the many. The process and objectives of technological advance are determined, nationally and internationally, by such factors. If we are to believe in the evolutionary concept of progress toward a better society and a better quality of life with freedom and justice, then the social and human aspects of development must be regarded as prerequisites. According to what W.F. Wertheim terms the emancipation principle, "emancipation from the forces of nature and emancipation from domination by privileged individuals or groups, therefore, go hand in hand to mark human progress."[5] Freedom and progress indeed constitute one and the same set of moral and spiritual values of development: human, social, and economic as well as scientific and technological.

In terms of scientific and technological advancement, all this means that the process and objective of change and development must be shifted from the all-too-familiar quantitative notion of growth to the qualitative one of freedom and justice. The world has indeed come a very long way from eighteenth- and nineteenth-century Europe, where an extremely high sense of optimism prevailed in the Age of Enlightenment. It was followed by the biological and social theories of evolution. That age was full of high and rising expectations of unlimited material growth on the one hand and social, cultural, and moral progress for mankind on the other. All this was to be achieved by means of science, technology, and industry.[6]

Obviously Europe's scientific and technological achievements also served as the moving force for its growing self-confidence and the conviction that soon developed into the hegemonic notion of the White Man's Burden. Hence all the expansionism and colonialism that followed and, along with it, all the hardships and alienation of both the displaced people within industrialized European society itself and the subjugated non-Western peoples the world over.

Thus the social and moral consequences of technological achievement were quite in contrast to what was optimistically expected before. At any rate, all the adverse phenomena serve to reveal the true nature, function, and results of science and technology. These still need to be objectively assessed and understood. As has been recognized, technology does not simply mean applied science culminating in an object, invention or even a mode of production – that is to say, something autonomous and neutral. As Johan Galtung describes it:

[Technology] carries with it a code of structures – economic, social, cultural and also cognitive. The economic code that inheres in Western Technology demands that industries be capital-intensive, research-intensive, organization-intensive and labour-extensive. On the social plane, the code creates a "centre" and a "periphery," thus perpetuating a structure of inequality. In the cultural arena, it sees the West as entrusted by destiny with the mission of casting the rest of the world in its own mould. In the cognitive field, it sees man as the master of nature, the vertical and individualistic relations between human beings as the normal and natural, and history as a linear movement of progress . . . [7]

In simple terms, it is subject to human and political decision with a view to authoritative allocation of values determining who gets what, when and how. And this structural relationship gives its own peculiar connotation to science and technology as a system of knowledge and its application. In principle, of course, science may be universally defined as a search for knowledge for its own sake. In reality, however, it is also part and parcel, and is in the service, of a particular socio-economic system. The interlocking between science and technology on the one hand and the socio-cultural context on the other gives rise to the need for what Susantha Goonatilake calls the "cognitive" mapping of physical and social reality.[8]

The truth is that even scientific knowledge continuously changes and develops over time. In its search for valid explanations of physical reality it not only operates under a specific world-view that constantly changes within the scientific community, but also interacts with the external socio-economic environment which also undergoes constant change. Scientific knowledge and theory therefore constitute a development process that is multidimensional and is bound up with, again in S. Goonatilake's words, "the internal social context within science, the external social context and the mode of production."[9] It is even more obviously so with regard to technology functioning as the applied side of science. As a matter of fact, it goes far beyond science to the stage of actual production and social relations. Science and technology have therefore to be conceived of objectively as a social phenomenon in themselves. And indeed, historically and empirically, the emergence and development of scientific technology have been recognized as an outgrowth of the interplay of extraneous economic, social, and political forces.[10] As a body of technical and practical knowledge it is, in the final analysis, subject to human volition, decision, and implementation.

The point to be made here is that all the adverse social and moral effects referred to above are not just a matter of self-interest or incidental abuse and misuse of technical knowledge. They are fundamentally a question of cultural and epistemological orientation that determines the state of mind and perception of "reality" as expressed in the current state of science and technology, both physical and social. Historically bound up with the rise of capitalism, modern science and technology have thus developed into an acquisitive and hegemonic scientific culture that sees itself as the absolute master not only of things but also over fellow human beings. Inherent in this cultural value system, needless to say, there is also a keen sense of historical mission to bring "progress" to the world at large with all its scientific and technological might, by force if necessary. This explains why the Industrial Revolution took place with so much ruthlessness and destructiveness, especially to rural and traditional community life. And out of these cultural centres the very same scientific world-view has also been brought to the non–Western world, where modernization and industrialization have been and still are forced upon it, with very much the same zealousness both from within and without.

RETHINKING "LIBERALISM"

Recognition of the human and social dimensions of technology can indeed be said to constitute a major step forward from the so–called classical model of human rights development.[11] The past three centuries have already witnessed the broadening of the human rights spectrum from the conventional set of civil and political rights and liberties to the newly claimed economic, social and cultural rights. Underlying all these combined negative and positive rights, it is to be noted, is the historical and empirical process of defining the status of man and his relationship to the state. At the beginning was the eighteenth-century notion of civil liberties, whereby the status of the individual was asserted as that of a self-sufficient and self-directing agent who needed little, if any, interference from the government. Government was then at best a necessary evil.

This view was to be followed and somewhat modified in the following century by way of a more positive concept of civil and political rights. Hence legal guarantees and enforcement on the part of the government came to be required for the attainment of equal rights of civic and political participation. Then finally came the demand for economic and social rights, involving the government's positive programmes of action to provide social welfare and to meet basic human needs, especially for disadvantaged groups of people in industrial society.[12]

Both the negative and positive aspects of human rights are obviously interrelated, representing libertarian and egalitarian streams of thought. However, as defined specifically within the framework of industrial capitalism, the libertarian aspect naturally and conventionally takes precedence over the egalitarian side. And this, more often than not, gives rise to socio-economic imbalances within the so-called liberal democracy itself. As John Strachey once observed, there is in capitalism – which is the historical moving force of modern liberalism – an "innate tendency to extreme and ever-growing inequality."[13]

To a large extent, the Western concept and practice of human rights is very far from being comprehensive and universal. As already noted elsewhere,[14] this conceptual partiality is inherent in the historical notion of natural law itself, which serves as the inspirational source of today's ideal and practice of human rights. According to John Locke, the father of liberalism, freedom simply meant being free to do what one liked. Indeed, as ideology, the natural law concept had its great historical achievement in opposing political absolutism and arbitrary rule and replacing divine right with the common man as the basis of political authority. But for all its broadening world-view, the then liberal idea was preoccupied first and foremost with the security and protection of property rights,[15] in the context of the rise of the middle classes in its time. In short, it was historically, and still remains, the liberalism of the haves, and this has grown into a force against the have-nots.

Within this conceptual framework of liberalism, at least two basic human rights still remain unsatisfied. First, externally, it is far from effective with regard to the third-world countries' most pertinent issues and problems of inequality. These are inherent in their agrarian socio-economic structure, and have been worsening in the course of modernization and industrialization. We shall revert to this topic later. Secondly, it is far from comprehensive with respect to the West's own pattern and process of industrialization, where the issues and problems of transition from the agricultural phase were simply taken for granted or ignored. Here the impact of modern scientific technology, among other things, loomed very large in bringing about rural dislocation and disruption.[16] But what was a loss in terms of rural human and social costs came to be viewed as a gain in terms of so-called economic growth and the proletarianizing of the rural sector, with the resulting benefit of cheap labour for the urban and industrial sector. Such was the price paid in the cause of so-called scientific, technological, and economic progress. Unfortunately, such is also the predominant trend of thinking and belief among the modernizing élites of most developing countries today.

For this very reason, so far as the issues and problems of human rights development are concerned, it is essential to look into the whole process of social change and transformation. It is in this sense that the social impact of technological growth and development needs to be re-examined and assessed. What happened in the Industrial Revolution of the West should be taken as a lesson to be learned and critically evaluated, instead of a model to be literally followed in the process of modernization of today's developing countries. This is the crux of the whole problématique of economic development in the third world. And this is precisely where the issues of human rights development and technological growth come to be interwoven.

In the eyes of third-world leaders, the logic of modernization requires accelerated economic development to be associated with industrialization and hence the adoption of modern Western technology, as if it were a ready-made solution. On the face of it, this sounds reasonable enough. But on account of the structural nature of Western technology, as well as the social and economic conditions of the third world, there are still quite a few questions to be clarified. In particular, how can industrialization be brought about in an overwhelmingly agrarian setting? And how can modern scientific technology be made use of with a minimum of human and social costs? Above all, how can technological development proceed while avoiding the pitfalls of dependence and subordination? All this involves a technological and structural change with direct implications for human rights, which had been historically bypassed in the Western experience but which today's developing countries must face.

HUMAN RIGHTS IMPLICATIONS FOR DEVELOPING COUNTRIES

At this point, the true nature of technological and industrial advance has to be set in a proper perspective of rights and obligations. In the name of civil and political rights and liberties, as historically derived from the concept of natural law, capitalism has made its way to the pinnacles of status and power. Yet in the course of its development, the economic and social rights of the majority of people have been trampled upon, thereby jeopardizing civil and political rights themselves. There thus comes about a powerful economic and technological force working towards domination and inequality. It was first set to work against its own rural people and labour, and then went on to overseas expansion, thus making itself economically as well as politically powerful and domineering. All this has been seen, incidentally, by capitalism and Marxism alike, as part of historical necessity and inevitability, at least in so far as the Industrial Revolution is concerned. At any rate, it is the empirical basis upon which the classical theory of economic growth has been established. The same can obviously be said of modern science and technology as generally conceived and practised up to the present time.

Of even more importance to the conception of human rights, which is particularly at issue here, is the people's potential and prospects for self-development, which have been suppressed and disrupted under hegemonic and exploitative regimes. The current capitalistic system and, for that matter, modern science and technology not only breed flagrant inequality within and among nations: they also see progress as a unilinear historical movement, that is, proceeding by stages as determined by capital and technology.

The alternative approach involves the far more fundamental question of cultural values and dynamism, through which science and technology can be made to contribute truly to human and social progress, together with technological advancement.[17] In contrast to the hegemonic and imposed industrial civilization currently perceived as uniform and universal, this approach gives full recognition to the diversity of cultures and values which, in P.C. de Lauwe's words, "guarantees constant renewal, dialogue and freedom of expression and is therefore the prerequisite for a truly democratic concept of community life."[18] It is mainly through respect for cultural pluralism and dynamism that the principle of equality and freedom can be secured and promoted along with economic and technological growth and development.

This point is most pertinent to and should be understood by today's developing nations, as latecomers in the field of modern science and technology. Within advanced industrial countries, hegemonic and exploitative relationships have been qualified and somewhat restrained within a democratic framework of civic and political participation. Many of the third world's developing countries, by contrast, are under authoritarian regimes and traditions and practically all the public decisions are left to the tiny groups of so-called modernizing élites.

In the past, colonial and semi-colonial countries and peoples were conquered and exploited as sources of raw materials and markets for manufactured goods. In the process, their traditional values and knowledge systems were transformed into a colonial culture that could not be much more than dependent and imitative.[19] With the passing of colonialism, there comes a new prototype of colonial culture, especially among the national élites, which looks to foreign capital and its accompanying science and technology as the agent of change and modernization. This type of modernization syndrome in turn serves as the dominant culture of the new ruling classes within the developing countries, thereby transforming these countries into dependent economies. History thus again comes full circle to the very same logic of industrialization and technological growth and development as some three centuries before, though in a new political–economic context. The difference is that in place of direct or indirect colonial rule, industrial capitalism has now transformed itself and developed into transnational corporations, with the third world's national élites serving as the point of contact in a context of dependent economic and cultural relationships. The hegemonic and exploitative structure of relationships remains basically the same, but is given an appearance of national identity and legitimate national aspirations and interests.

Notwithstanding all the nationalistic claims, however, the fact remains that these national élites' aspirations and goals are closely associated with and strongly inclined toward the Western master culture.[20] Here the cultural impact and influence of Western-style education and professional training has to be noted. This has been going on ever since colonial days, and it even intensified after the Second World War, with the coming of national independence. Nor is that all. The same kind of learning process and knowledge system has also been built into the so-called national education imposed upon the population at large. In short, as S. Goonatilake again has observed, what emerged and developed as hegemonic scientific culture has been carried over from the Western centres and then assumed to be legitimate for non–Western societies.[21]

It is through such socio-cultural processes and conditioning that modernization and modern scientific technology become the transmitter of hegemonic social relations, within and among nations. Hence, modernization and industrialization have come to be associated exclusively with the capitalistic process of growth. Thus emerge the enclaves of modernity initially set up for the purpose of import substitution and manufacturing self-sufficiency, but currently heavily geared to export processing for the so-called free-world market. Capital-intensive technology and attendant technical know-how have been and still need to be imported, thereby bringing about production relations that are under external capital and technological as well as market control. All this is hardly, if at all, in consonance with the rest of the domestic economy. And all of this is in the name of growth, with the hope that the material benefits thus derived will somehow trickle down to the underprivileged sectors of the population. Meanwhile, for at least three decades now, this unbalanced growth strategy with its "innate tendency to extreme and ever-growing inequality" has increasingly expressed itself in the extreme form of glaring and growing poverty and unemployment as well as chronic indebtedness among developing nations. And all this has been going on, ironically, in the midst of manifest affluence not only among the industrially advanced nations but also in the modern sectors within developing countries themselves.

Needless to say, the incidence of all these ill effects is bound to bear heavily on the rural and traditional sector. Underlying the phenomenon of social and economic ills is the state of their economic and technological backwardness vis-à-vis the modern urban sector. In the process, a dualistic and unequal structure has been created within society. It is in this light that the existing state and the future prospects for human rights in developing countries should be understood and assessed. The implications obviously go far beyond the issues between North and South. They certainly involve more than the conventional set of human rights, as developed from the standpoint of the rising mercantile and industrial capitalist classes within the cultural context of the industrial West.

If the Western historical experience is to be any guide at all the issues have to be traced back to the plight of those in the rural sectors who were forcibly dislocated and alienated in the process of technological advancement and industrialization. So also must they be attributed to the plight of the overwhelming majority in the rural and traditional sectors of today's developing countries. Moreover, in addition to the adverse impact on economic and social rights and civil and political rights, their traditional cultures and productive capability as a means of self-expression and creativity are being suppressed and disrupted. Not only are they deprived of the benefits of modern scientific technology, but their own cultural potential for self-development also comes to a standstill and eventually falls into abeyance. Under such structural constraints, modern science and technology per se can be no compensation for the cultural deprivation thus caused to the common people. It only serves, and is preserved, as the exclusive domain of the élite for the maintenance of their privileged status and domination over the rest of society.

TOWARDS SELF-RELIANCE FOR CULTURAL FREEDOM AND CREATIVITY

The observations made concerning the structural nature of modern science and technology by no means suggest an anti-Western philosophy or a policy which opposes modern scientific knowledge and its application. Neither do they imply a need or a desirability to fall back on the traditional past and to keep away from the realities of the contemporary world. That would be tantamount to compromising one's own cultural and creative potential for contributing to progress – a prerequisite for the quality of life and, even more, for freedom and creativity. Besides, life in today's world involves ever-increasing interdependence, and interrelationships in society, both national and international, are becoming ever more frequent, more intensive and more penetrating. On their part, modern science and technology have definitely come to stay, whether one likes it or not, and they will stay as world science and world technology. Significantly, too, they are both accessible and available for creative and positive use. All this is a fact of life that one can ignore only to one's own detriment.

The solution sought by the developing countries ought therefore to be a more positive and constructive one. The answer certainly does not lie in either escapism or aversion to scientific knowledge and technology as such. It is fundamentally a question of how non-colonial science and creativity can be promoted and developed, in order that real human and social progress can be achieved. This only means that ways and means have to be found for modern scientific technology to be made use of, not for purposes of domination, but as a liberating tool, and thus for transforming all the productive forces available to society into a balanced and self-sustaining process of growth and development.

In development literature, this line of approach has generally been referred to as self-reliance. This is obviously the most logical alternative to the current situation of technological hegemony and dependency. As a matter of principle, it sounds readily acceptable. However, when it comes to actual practice, whether in the realm of public policy or in research conceptualization and methodology, the purely technical aspects assume such importance that the concept of self-reliance is whittled down to merely serving so-called national and professional aspirations and supporting whatever reinforces power and prestige. Amidst such élitist attitudes, exclusive concern will be directed to such subjects as high technologies, advanced R&D and manpower, and advanced technical training, all of which are treated as ends in themselves. It should not be difficult to see that this line of approach also spells forms of domination/subordination relationships that one is trying to avoid in the first place.

Of more relevance to the human rights issue under discussion, the élitist tendency also serves as the cultural basis of complementary attitudes on the part of the dominating and the dominated, on the basis of which the process of technological domination operates. As P.C. de Lauwe observes of the existing state of affairs: "On the side of the (dominating), the temptation to use scientific and technological superiority is all the greater inasmuch as it enables them to beat trade competitors and extend their political influence, while at the same time ensuring their national defence. As for the dominated, the loss of initiative, freedom and creation that goes together with over-rapid changes imposed from abroad is made acceptable by the material benefits that modernization brings."[22] In this fashion, technological, economic, cultural, and military domination have come to be intertwined.

It is therefore important to be aware of another level of hegemonic scientific culture which comes in disguise and in the name of nation-building and national progress. Indeed, the world has gone through the stage of national self-determination, only to end up with a new breed of domination and oppression within nations. If there is to be any hope at all for progress with respect to human rights, this is the time to give due recognition to the needs and aspirations of men and women at the grass roots.

Mention has earlier been made of the issues and problems of deepening inequality and poverty in the third world. Again, the solution is not simply a question of providing material well-being. Nor is it a question of improving the distribution of income or material benefits by the powers that be, however well-intentioned or benevolent. It is the fundamental problem of a lack of appropriate productive capability whereby one can fully develop oneself in society. This lack of productive capability, and therefore of self-reliant development, is, it must be strongly emphasized, to be seen not simply as an individual or incidental phenomenon but essentially as collective and systematic in character. There can be no denying the fact that, because of the unbalanced process of economic growth imposed by the national élites in co-operation with the global powers, industrialization has been forced and accelerated to the detriment and disintegration of the rural and agricultural sectors. In consequence, the peasantry of the third world has become deprived of the productive and innovative capability which would enable it to hold its own. In short, it is reduced to becoming technologically backward as well as economically and politically subject to domination and manipulation.

Fundamentally, the question of self-reliance in science and technology is concerned with the cultural freedom and creativity that have been lost in the process of forced industrialization. Ironically enough, both capitalism and communism, though ideologically poles apart, pose quite a comparable problématique here. In his critique of the spiritual loss during the era of collectivization in the Soviet Union, Andrei Sakharov expresses the hope that the earlier spirit that gave life its inner meaning "will be regenerated if suitable conditions arise."[23] The same problématique could also be said to arise in connection with spiritual and cultural values under the current form of capitalism. In fact, by the very same logic of technological domination, the two, as agents of industrialism under the Second Wave civilization, are not much different.[24] Each could be said to represent the consequences of its respective historical factors and conditions. The point is that neither of them can provide the answer to the question of cultural freedom if carried to extremes, as has so far been the case.

In terms of the right to development, specific attention has to be given to the rural and agricultural sector that has been neglected, even oppressed, for so long. This line of approach is most pertinent to today's developing countries, which have an agrarian background. The fact is that no agrarian societies have ever been without technical knowledge and inventiveness. They have their own traditional means of learning and skills in technological adaptation and innovation. Moreover, these traditional values and technologies do not exist in a vacuum. Underlying them is local and endogenous wisdom and a creative learning process that has been accumulated over generations. For all their seemingly non-scientific attributes, they are directly related to people's real and relevant needs and organizational and environmental conditions.[25] And, most important of all, they are expressed through free will and with a rationale of their own. Besides, for all their tradition-bound nature, the peasants themselves are actually quite receptive to new and modern technologies introduced from outside whenever they are relevant and feasible and demonstrated to be so in practice.[26] This clearly points to the value and dynamism of traditional knowledge and creativity. However, under the existing dualistic structure, they are being left behind and allowed no chance of gaining the benefits of modern science and technology.

All this, of course, is not to be taken as a purely romantic vision. On the contrary, the intellectual limitations and constraints of rural environments have also to be recognized for what they are, especially in the face of a changing world. There is, however, a real need to ensure that receptivity and adaptability to modern scientific knowledge and technology do not result in social disruption. Again, from the standpoint of human and social progress, the modern and the traditional have to be looked at and acted upon as complementary to each other, not as being poles apart as has been the case up to now. This criterion of complementarity would naturally raise the question of rights and obligations between the haves and the have-nots in such problem areas as, for instance, the right of technological choice and the right of access to technological information both within and among nations. It would also raise the question of the feasibility and desirability of a new international technological order, which has been so much talked about lately. More crucial to the issue of technological self-reliance, however, is the matter of socio-cultural adjustments within the developing countries themselves.

First of all, the state of endogenous technological knowledge and experience needs to be carefully reviewed and evaluated. Furthermore, to set each country's comparative advantages in a proper perspective, an extensive investigation has to be made of the basic human and natural resources and endowments within the country and the socio-cultural and organizational characteristics of both the agricultural and industrial sectors. There may be, for example, countries with an abundance of labour, or those with abundant natural resources. Or, again, some may have mixed physical advantages and some may have more or less advanced industrial bases. All such potential or comparative advantages would constitute an empirical basis upon which to review and set out the overall direction and pattern of technological growth and development.

In any case, the most decisive factor rests, in the final analysis, with human resource development itself. For all the material and physical nature of technology, technological change and development means in essence modifying and transforming the productive forces in society. In contrast with the existing partial view of growth, the objective and principle of self-reliance takes a holistic view of human and technological development. In terms of human resource development for the purpose of technological self-reliance, this means in effect that rural human resources must be looked upon not merely as a production input, as many an expert has made them out to be, but essentially as consisting of creative beings capable of self-reliance and self-development. This is precisely what was implicit in the concept and principle of the natural and equal rights of man, but which, as mentioned earlier, has been negated under the impact of hegemonic scientific and industrial culture.

CONDITIONS OF SELF-RELIANT DEVELOPMENT: ASIAN CASE-STUDIES

Now that self-reliance in science and technology has been identified with cultural freedom and progress, and thus presents itself as an alternative to the existing state of domination, dependence and underdevelopment, the further question arises as to how it could be brought about. As earlier emphasized, science and technology are not in themselves at issue here. But as part and parcel of the social system and of a social process, they are bound to socio-cultural factors and conditions as they exist. The repeated call for a change in attitudes on the part of both the dominating and the dominated[27] will most likely come to naught without infrastructural change, or what Sakharov looks forward to as "suitable conditions." It would be beyond this paper's scope to suggest how those conditions could be brought about, and by whom and by what forces. This has to be left to the actual and specific socio-political actors and forces involved. However, a combination of factors and conditions could possibly be identified here as guidelines for action and implementation.

For this purpose, specific reference should be made to a recent study under the aegis of the United Nations University, on "Self-reliance in Science and Technology for National Development" (to be referred to as the Self-reliance Study), involving six selected Asian countries: Philippines, Thailand, India, China, the Republic of Korea, and Japan.[28] Here the focus on infrastructural aspects also serves as the common line of approach. The relationship and interaction between technology and society are first examined, and then the concept of technological self-reliance is broadly defined. The study's main objective is to make empirical inquiries into the socio–cultural factors and conditions that may have a bearing on the possibility of promoting a self-sustained process of technological growth and development in various specific social contexts. In addition to the specificity to be expected of each country study, a broader regional perspective of self-reliance is also implicit in the range of variations expressed in terms of stages of develop- ment: i.e. from the highly advanced industrial Japan to the newly industrialized Republic of Korea and India, and through to the other developing countries with a largely agrarian background.

Obviously, the six selected countries cannot be regarded as representative of the whole Asian region. Besides, those characterized as agrarian and traditional Asian societies have been undergoing varying degrees of modernization, not to speak of the vast differences in their geographical, economic, cultural, environ- mental, and historical backgrounds. Nevertheless, out of all these specificities and diversities, the whole issue of self-reliance and human rights could be perceived from the common standpoint of agrarian and traditional societies. In the light of its agrarian and traditional past, indeed, Japan can be said to share a certain Asian perspective. In fact, in an earlier study on the Japanese experience, specific atten- tion was clearly given to the infrastructural aspects of technology, "in the context of self-reliant efforts toward development of developing countries."[29]

On the basis of this shared development perspective, the concept of self- reliance is not to be understood in the negative and simplistic sense of aiming at autarchy. Its objective is not for each country or nation to seek self-sufficiency exclusive of one another. On the contrary, ways and means should be found for international exchange and co-operation to take place in a creative and construc- tive manner on the basis of egalitarian relationships. Self-reliance in science and technology is thus to be perceived positively, in the collective sense of cultural creativity and interdependence.[30] It also follows that within each country itself, self-reliance needs to be conceived of in holistic and dynamic terms as social capacity to innovate and adapt existing and new technologies. And this must be understood in a socially integrated sense, and not confined to any particular tech- nology or sectors of the economy.[31] Finally, in concrete terms, this criterion of "social capacity" has to be related to at least the following three basic require- ments:

1. Optimal use of local resources and the meeting of basic needs.
2. Development of related indigenous manpower and human resources.
3. Development of grass-roots institutions and participation along the path of national development.[32]

All these requirements of technological self-reliance are of course interrelated. But the last component has a most pertinent bearing on the issue of human rights. Mention has earlier been made of the unbalanced growth strategy and consequent rural underdevelopment and poverty within the developing countries of Asia. During the past decade, increasing attention has been given to rural needs and problems. Within the existing framework of so-called rural development policy, however, it still remains fundamentally top-down in approach. Hence current government concern and attempts to deal with the critical problems of income distribution and social welfare. This is all very well. But it remains benevolent at best, which is very far from the real objective of promoting self-reliance. Industrial and agricultural disparity continues to be taken for granted. Industry remains the topmost priority and any hope of improving rural socio-economic conditions is made to hinge on further industrial expansion. In short, the industrial sector remains the sole answer as the source of employment and the non-farm source of rural income.[33]

In truth, the real solution lies in the technological and productive capability within rural communities themselves. As it happens, the rural sector has been suffering not merely from maldistribution of the benefits of growth. That is just the inevitable consequence of the adverse impact of modern science and technology under hegemonic industrial culture, whereby rural manpower and traditional technology as the basis of cultural creativity have been alienated and disrupted. All this is of course man-made and concerned largely with the question of political economy to be touched upon shortly. What needs to be stressed at this point is the question of reviving and regenerating this potential creativity, not for its own sake, but to serve as the basis on which modern scientific technology could be adapted and made use of. And here rural participation in technological growth and development is essential, so that the choice and assessment of technology, instead of being imposed or forced upon them, can effectively be made within and by the rural communities themselves. Grass-roots participation is obviously a most meaningful way of mobilizing endogenous resources in the process of long-term growth and development, and should in the last analysis have significant implications for developing technological and productive capability, and, indeed, for income redistribution.[34]

This is not all. Mention has already been made of the peasantry's receptivity to modern innovations. It is obvious that this receptivity by itself, no matter how many development services are provided, cannot really be effective unless the peasants are appropriately educated and trained.[35] This is indeed essential for a self-sustained process of technological growth and development. Here the conventional concept of the social and cultural right to education, in particular, needs to be briefly re-examined to get a clear perspective of the whole problem.

Under the existing so-called national education system in developing countries, the rural population could be said to have been deprived of their natural and equal rights. According to the Universal Declaration of Human Rights, the right to education is broadly defined as full development of human personality and a sense of dignity. But the actual learning process and the objectives of education are by and large imposed in conformity with "national" requirements that are in effect élitist-oriented in spirit and drawn heavily, in not entirely, from exogenous and hegemonic sources. It is not only irrelevant to basic needs in rural environments, but it also deprives the whole rural community of its own human resources and therefore its potential for self-sustained technological growth.[36] The right to education is not just the right to any kind of education. It is fundamentally a question of an appropriate education that would form the basis for developing appropriate technological capability. The adverse impact of miseducation on both individuals and rural communities indeed points to a serious shortcoming in the current scheme of thought concerning human rights as a whole. It also has significant implications for the problématique of self-reliance in a collective sense.

The point made here is that the educational system, among other things, needs to be reformed as the infrastructure for advancement toward technological self-reliance and cultural freedom and creativity. This, as emphasized earlier, is a question of giving due recognition and respect to cultural pluralism and dynamism. The rural sector too is in need of its own educated and scientifically innovative manpower, no less than the urban and industrial sector. The need is even greater in that the rural sector has been subordinated and exploited in the long process of deliberate, unbalanced economic growth. The urban–rural dichotomy, repeatedly referred to, does not mean that each should go its own way, or that the rural sector has to remain agricultural forever. Technological self-reliance should in the long run enable the rural sector itself to be industrialized and even cope with the higher technologies available. But all this should be on the basis of concrete comparative advantages, material, human, and cultural. In particular, it has to be a process of growth, technological or otherwise, wherein people at the grass roots are able and willing to participate fully and meaningfully.

The basic socio-cultural factors required for technological self-reliance would be next to impossible to achieve without favourable politico-economic conditions. After centuries of dependence and underdevelopment, these conditions also need to be revitalized. The first condition is concerned with the lack of autonomy in the decision-making process.[37] This is typical of socio-economic change and transformation being externally forced upon the traditional societies of Asia. The importance of restoring full national sovereignty ever since the Meiji era is alluded to in the Japanese Experience. In the Self-reliance Study referred to above, this particular requirement has been expressed in a variety of ways. The India Report, for instance, points to domestic capital formation as a necessary condition for economic and technological self-reliance. The China Report looks to the historical perspective of "basic completion of socialist transformation" as a prerequisite for the launching of meaningful self-reliant development. The Philippines Report presents a classic case of modernizing élitism and authoritarianism in traditional societies in Asia. Immediately after the end of colonial rule, its political economy came under the domination and control of transnational corporations, backed up by the nation's élites and by corruption. This state of affairs has had a negative effect on the nation's capacity for decision-making and choice of strategy and technology for national development. In other words, along with transnationalization of economic relations, the national leadership has in the process also become transnationalized in its values and outlook, and even in its loyalty.

Closely related to the need for autonomy in decision-making are what Dieter Ernst terms "key developmental objectives," in which, among other things, agriculture plays an essential part.[38] This is logical enough in view of the agriculture-based nature of Asian economies. As mentioned earlier, the policy focus on agricultural development should not bar rural economies from becoming industrialized. But there is a vast difference between industrialization and the progress or stagnation of agriculture. Here is the crucial policy choice between balanced as against unbalanced growth, which also involves the direction and pattern of technological development. This is implicit in the Thailand Report, which provides an exclusive focus on the technological problématique of agricultural production, as well as the feasibility of a bottom–up strategy of rural development. Indeed, in terms of the socio–economic realities of the developing countries, the modernization of agriculture has its own value in enhancing the main source of food supply and employment, and, significantly for long-term industrialization, in creating an effective domestic demand for industrial goods.[39]

Finally, at regional and global levels, the objective of human and social progress requires that the existing hegemonic scientific culture should give way to collaborative relationships. On the part of the advanced nations, there is yet to be a change of heart, notwithstanding all the talk about a New International Order. From the developing countries' standpoint, the need for interdependence should be self–evident. In default they are likely to be doomed to stagnation in the midst of a changing world. Even so, there is also good reason for caution in the matter. Interestingly enough, the possibility has been raised in the Japan Report of creating a linkage between industrial and agricultural technologies. In the light of the need for a collaborative relationship between advanced industrial Japan and developing Asian countries, agro–related technologies have been suggested for consideration. But on account of the long-standing unequal international division of labour, and thus the existence of dependent relationships, strong negative reservations have been voiced.[40] To a great extent, this represents a very realistic reaction, in view of the existing state of political economy and science and technology within the developing countries themselves.

With the feasibility of international interdependence in mind, a good starting-point would be in promoting a process of collaboration and exchange of information about scientific and technological research and educational and training facilities.[41] This is easier said than done. In the first place, the developing countries themselves must be sufficiently prepared for the task involved. Again, the importance of strengthening the endogenous base of technological growth and development cannot be overemphasized. In the second place, there is the long-standing problem of intellectual property rights that all too often has been allowed to stand in the way of effective co-operation. There is of course a legitimate concern that these rights be preserved and protected. However, a pertinent question can also be raised as to the distinction to be made between science and technology as a body of knowledge for social benefit as opposed to that used for commercial purposes. Should intellectual property rights be treated as absolute? If so, that simply means going back to square one. It would result in continuing movement along the path of old-time mercantilism and industrialism, at great cost to humanity. It is likely to become even more threatening now that the world at large is moving into the information age. And this poses a critical question for all to answer, and a critical choice that must be made. A breakthrough would open up new horizons that would go a long way along the path of enlightened self-interest and, therefore, of creativity and progress.

CONCLUDING REMARKS

So far as developing countries are concerned, the foregoing discussion comes down to one most fundamental question – whether, and if so how, self-reliance is to be recognized as a right associated with cultural freedom and the capacity to grow and develop. Again, implicit in this is equal respect for cultural pluralism and dynamism. This goes far beyond the conventional libertarian or egalitarian approach to the problem of human and social relations. It is of course pointless to get stuck in the historical past. But development efforts toward self-reliance also involve restoring and regenerating the endogenous creativity that has been lost under the impact of industrial scientific culture.

This adds a cultural and thus a collective dimension to the problem of technological self-reliance – one that goes, again, beyond the mere question of the individual's right to "enjoy the benefits of scientific progress and its application."[42] It is basically concerned with the problématique of the cultural identity of whole rural and traditional communities that have been undergoing adverse social change and transformation. This is by no means a defence of traditionalism; however, there is no valid reason for allowing the current trend for hegemonic industrial culture to go on oppressing people for its own sake. There is no such thing as historical necessity or inevitability, as assumed by both capitalism and Marxism. The real and most obvious course is to let endogenous sources of knowledge and creativity be revitalized and developed as the basis upon which modern scientific technology could be effectively adapted and assimilated.

All this means that there should be no inherent incompatibility between modern and traditional technology. The contradiction has only been man-made, historically speaking, and the path of future development can be changed for the better by human intervention. In developmental terms, traditional technology needs to be upgraded to modernized intermediate technology. And in this perspective, it should be in a symbiotic relationship with exogenous sources of scientific knowledge and technology. Modern scientific technology therefore has always a great role to play, not to supplant, but to supplement indigenous technology. In short, the modernization and industrializiation of developing countries could and should take their own respective routes.

For all its feasibility, technological self-reliance and cultural freedom is in the final analysis a question of political relationships, both within and among nations. Like all the other human rights problématiques, it requires structural change and transformation. In this sense, it is likely to remain an open question for quite some time to come; that is, in the absence, in Fouad Ajami's words, of "the politics of love and compassion" as against the current "realism,"[43] according to which the sole objective of power is to rule.

NOTES

1. Friedrich Rapp, "Philosophy of Technology," *Contemporary Philosophy: A New Survey*, vol. 2 (Marinus Nijhoff The Hague/Boston/London, 1982), pp. 376–378, in reference to H. Rumpf, G. Simondon, and A.G. van Melsen.
2. A. Rahman, "The Interaction between Science, Technology and Society: Historical and Comparative Perspective," *International Social Science Journal*, vol. 33, no. 3 (1981): 508–518. For a more detailed discussion of the impact of European science on South Asia, see Susantha Goonatilake, *Aborted Discovery: Science and Creativity in the Third World* (Zed Books Ltd, London, 1984), chaps. 3 and 5.
3. Robert S. Cohen, "Science and Technology in Global Perspective," *International Social Science Journal*, vol. 34, no. 1. (1982): 63.
4. Cohen (note 3 above), p. 62.
5. W.F. Wertheim, *Evolution and Revolution: The Rising Wave of Emancipation* (Penguin Books, Harmondsworth, 1974), p. 47. See also pp. 35–48 for elucidation of the criterion of progress and the emancipation principle.
6. Rapp (note 1 above), p. 361; Celso Furtado, "Development," *International Social Science Journal*, vol. 29, no. 4 (1977): 628–629.
7. Johan Galtung, "Towards a New International Technological Order," In *North/South Debate: Technology, Basic Human Needs and the New International Order*, Working Paper no. 12, World Order Models Project (Institute for World Order, New York, 1980), p. 4.
8. Goonatilake (note 2 above), pp. 1–2.
9. Goonatilake (note 2 above), pp. 1 and 80–81. As stressed here, the search for explanation of physical reality involves not only physical science but also social science.
10. Cohen (note 3 above); Peter Grootings, ed., *Technology and Work: East–West Comparison* (Croom Helm, London/Sydney/New Hampshire, 1986), pp. 277–283.
11. Richard P. Claude, ed., *Comparative Human Rights* (Johns Hopkins University Press, Baltimore/London, 1976), p. 7.
12. Claude (note 11 above), pp. 41–42.
13. Quoted in T.B. Bottomore, *Elites and Society* (Basic Books, New York, 1964), p. 34.
14. Saneh Chamarik, "Buddhism and Human Rights," *Human Rights Teaching, Unesco Biannual Bulletin*, vol. 2, no. 1.(1981): 16. On the point of the conceptual shortcomings of liberalism, see C.G. Weeramantry, *Equality and Freedom: Some Third World Perspectives* (Hansa Publishers Limited, Colombo, 1976), p. 10; and Fouad Ajami, "Human Rights and World Order Politics," World Order Model Project, Occasional Paper no. 4 (Institute of World Order, New York, 1978), pp. 2–4.
15. John Locke, *Two Treatises of Civil Government*, reprinted by J.M. Dent & Sons Ltd (London, 1953), pp. 119 and 129–141.
16. Cohen (note 3 above); Furtado (note 6 above), pp. 630–631
17. Paul-Henry Chombart de Lauwe, "Technological Domination and Cultural Dynamism," *International Social Science Journal*, vol. 38, no. 1 (1986): 105–109.
18. Chombart de Lauwe (note 17 above), p. 107.
19. Goonatilake (note 2 above), pp. 91–114.
20. A. Rahman (note 2 above), p. 520.
21. Goonatilake (note 2 above).
22. Chombart de Lauwe (note 17 above), p. 108.

23. Referred to in Donald Wilhelm, *Creative Alternatives to Communism: Guidelines for Tomorrow's World* (Macmillan, London, 1977), p. 84.
24. Alvin Toffler, *The Third Wave* (Bantam Books, 1981), pp. 99–102.
25. Goonatilake (note 2 above), pp. 114–116.
26. For instance, Thamrong Prempridi et al., *Self-Reliance in Science and Technology for National Development: Case Study of Thailand*, Final Research Report submitted to the United Nations University, Tokyo, December 1986, pp. 86–90. In Thailand, with which this writer is particularly familiar, it is precisely this aspect of local wisdom and potential creativity that attracts specific attention among NGO volunteers after their long experience with rural development work. There are no doubt similar trends of interest in other developing countries.
27. For instance, P.H. Chombart de Lauwe (note 17 above), pp. 108–109.
28. The six country studies are to be referred to in this paper respectively as the Philippines Report, Thailand Report, China Report, India Report, Korea Report, and Japan Report.
29. Takeshi Hayashi, *Project on Technology Transfer, Transformation and Development: The Japanese Experience* (United Nations University, Tokyo, 1984), pp. 2–3.
30. Enrique Oteiza and Francisco Sercovich, "Collective Self-Reliance: Selected Issues," *International Social Science Journal*, vol. 28, no. 4 (1976): 666–667. See also Dieter Ernst, "Technology Policy for Self-Reliance: Some Major Issues," *International Social Science Journal*, vol. 33, no. 3 (1981), p. 476.
31. Report on the Workshop on Self-reliance in Science and Technology, The United Nations University, Tokyo, 31 October–3 November 1984, p. 9. See also Ernst (note 30 above), p. 467.
32. Ernst (note 30 above), pp. 467–471.
33. As reflected, for example, in the Korea Report, pp. 16–25.
34. Ernst (note 30 above), p. 468.
35. Chombart de Lauwe (note 17 above), pp. 109–111.
36. For the dependent state of science education, both physical and social, see Goonatilake (note 2 above), pp. 97–116.
37. Ernst (note 30 above), p. 467.
38. Ernst (note 30 above), p. 469.
39. Christer Gunnarsson, "Development Theory and Third World Industrialisation," *Journal of Contemporary Asia*, vol. 15, no. 2 (1985): 198–200.
40. The United Nations University's workshop report on Self-reliance in Science and Technology, Beijing, 6–10 October 1985.
41. Oteiza and Sercovich (note 30 above), p. 667.
42. International Covenant on Economic, Social and Cultural Rights, Article 15(b).
43. Ajami (note 14 above).

BIBLIOGRAPHY

Claude, Richard P., ed. *Comparative Human Rights*. Johns Hopkins University Press, Baltimore/London, 1976.

Goonatilake, Susantha. *Aborted Discovery: Science and Creativity in the Third World*. Zed Books Ltd, London, 1984.

Grootings, Peter, ed. *Technology and Work: East–West Comparison.* Croom Helm, London/ Sydney/New Hampshire, 1986.

Rapp, Friedrich. Philosophy of Technology. *Contemporary Philosophy: A New Survey.* Vol. 2. Marinus Nijhoff, The Hague/Boston/London, 1982.

Weeramantry, C.G. *Equality and Freedom: Some Third World Perspectives.* Hansa Publishers Ltd, Colombo, 1976.

Wertheim, W.F. *Evolution and Revolution: The Rising Wave of Emancipation.* Penguin Books, Harmondsworth, 1974.

4

Human Rights and Scientific and Technological Progress:
A Western Perspective

TOM J. FARER

My mandate, as I understood it, was to sketch the principal orientation of peoples in the capitalist democracies of North America and Europe to the human rights issues implicated in the constantly growing capacity of men and women to manipulate the natural world and to influence virtually every aspect of human life in ways hardly imagined just a few decades ago.

TENSION BETWEEN THE SCIENCES AND THE HUMANITIES

The juxtaposition of "science and technology" with "human rights" in the overall project description implies, I believe, a felt tension between two ways of knowing the world, between two distinct yet constantly interactive realms of knowledge: the realms of science and the humanities. The quickstep of science and technology increases exponentially the means for conscious intervention in everyone else's personal and social life by the small minority of people possessed of the requisite knowledge, capital, authority, and/or coercive power. In response, morally sensitive members of the human community, including some who are themselves positioned to exploit new knowledge, search desperately for standards to channel evolving technologies toward serving rather than subverting broadly shared interests.

Perhaps their search is driven by an even deeper concern. The leaps of scientific and technological knowledge threaten to do more than sharpen the pitch of extant hierarchies and increase the destructive potential of conflicts between élites. The new knowledge imperils our very sense of what it means to be human: our subjective feeling of responsibility; our belief in the capacity for moral action and personal improvement.

Our sense of what it means to be human depends on our conviction, however unconscious, that there exists a zone of autonomous, self-conscious choice, constricted but never entirely occupied by chance and genes and chemistry, and defensible, albeit not always successfully, against the intrusion of state and private power. It depends also on an ingrained sense of what is "natural" or "authentic" and immanent in the human condition as opposed to what is accidental, transitory, and fabricated.

The humanities are a collective record of our species' claims about itself, of its deepest beliefs, of its inner life, its consciousness of the freedom to create and to dream. They are a declaration of uniqueness, a dictionary of meaning, a thread of continuity. And they are, therefore, a statement of the values integral to the very idea of themselves human beings have evolved and nourished and must sustain to prevent an ineffable, indeed unimaginable, loss of coherence.

THE IDEA OF HUMAN RIGHTS

The idea of human rights is, on the international plane of existence, the formal normative expression of those values. And like them it is a blend of two moral traditions that, in one guise or another, have competed and co-operated through the whole course of Western history. One tradition, encapsulated in the "principle of utility," declares that our pre-eminent moral test must be the relative capacity of proposed actions to maximize the welfare or happiness of the community.[1] This is not a collectivist approach. The community is not treated as an organism distinct from its individual members, a whole greater than the sum of its parts. Community welfare is simply the sum total of the felt welfare of each community member. Thus if many benefit greatly from a proposed act and a few are injured and no alternative course would maintain a comparable range and intensity of benefits while reducing the incidence of injury, the act is morally justified according to the principle of utility.

Sometimes opposed in practice and always in theory is the moral tradition emphasizing the individual as the possessor of certain inalienable rights, rights which he or she may lose by unjustly diminishing the rights of others, which can be voluntarily waived, but which cannot be stripped by a majority however great its size or exigent its need.[2]

This distinction is relevant to my efforts at delimiting the issues I have been asked in very general terms to address. At its inception, the global human rights movement emphasized political and civil rights, rights of an essentially individual character, rights which can to a large degree be realized merely through the inaction, i.e. the tolerance, of governments. It was not long, however, before many advocates of a non-Darwinian politics demanded equal status for economic and social rights. Their realization required – in most if not all cases – not merely tolerance by the state and enforcement of that public order that incidentally protects the individual from at least the cruder forms of private coercion; their realization required a distinct, often a novel, allocation and quantity of public expenditure and regulation. Unlike such core political rights as free speech, economic and social rights deal with finite goods: more for some members of the community not infrequently means less for others, particularly in the short run. Partially as a consequence of that fact, economic and social rights have remained controversial. For the generality of the globe's inhabitants, they have also remained very far removed from reality.

"THIRD-GENERATION RIGHTS"

In recent years, some UN members and certain non-governmental organizations have advocated recognition and implementation of what are sometimes called "third-generation rights," such as the right to peace, to development, and to a healthy environment.[3] Although these goals are widely shared in the abstract, even some adherents of the human rights movement in the high-tech capitalist democracies have openly doubted the utility of expressing them in the language of human rights lest the latter lose their now well-defined contours and their claim, at least in the case of political and civil rights, to immediate implementation.[4]

But there are other reasons for the resistance to the campaign on behalf of third-generation rights, a resistance which, not accidentally, is led by governments and intellectuals from the industrially advanced capitalist states. One is the belief that the proclaimed "right to development," rather than expressing a felt concern for universally relevant human values, is simply a tactic employed by the third-world bloc at the United Nations as part of a campaign to transfer wealth from the North to the South.[5]

Even assuming the accuracy of the accusation that the "right to development" is designed to or would perforce require such a transfer (rather than requiring changes in the structure of international economic relations that would promote economic growth in the North as well as the South),[6] it does not necessarily follow that the claimed "right" is simply a moral veil for crudely political demands. On the contrary, if the operational substance of the "right" is a claim on the affluent for assistance in reducing third-world poverty, the right has morally appealing content.

Attempts to pre-empt that latent appeal assume the form of two propositions: first, that the "right" is intended to effect a transfer of resources from Northern taxpayers (including many persons who are not affluent) to Southern governing élites (who are); secondly, that the principal causes of Southern poverty are the policies of Southern governments (rather than unfair centre–periphery relationships stemming from the era of colonial servitude). Two corollaries of the second proposition are that any transfer will not relieve third-world misery, since its root causes will endure, and that the South cannot base any claim against the North on grounds of the latter's supposed unjust enrichment at the former's expense.[7]

A second objection from some Western sources to third-generation rights, particularly those to development and a healthy environment, is that like the second generation they imply a heightened level of activity by the state and, therefore, an expansion of its power. Under any circumstances, it is argued, expanded state power has problematic consequences for the classical liberal rights clustered around the notion of a broad zone of individual freedom from state regulation. Moreover, if, as some believe, statist intervention in free markets has tended to inhibit growth and concentrate income in the third world, any declaration of rights encouraging intervention will degrade the welfare of third-world peoples.

Right-wing Western polemicists often represent North–South issues as expressions of a fundamental cleavage over the centrality of human freedom; Northern governments (above all the United States) are pictured resisting Southern efforts to subordinate the individual to the state in the name of collective interests. The papers of my colleagues who have attempted to present a distinctively third-world view of the science–technology–human rights problématique (particularly those of Drs Herrera, Weeramantry, and Chamarik) demonstrate the inaccuracy of this polemical dichotomy. For it is evident that concern for individual autonomy stands at the heart of their work. Their differences with American conservatives turn, then, not on conflicting views about the relative importance of individual and communal interests, but rather on the necessary means for expanding, and the main threats to, the realm of human freedom.

Conservatives appear to assume that electoral competition and an economy marked by very limited state involvement are both necessary and sufficient conditions for the fullest possible realization of personal autonomy. Third-world scholars like my colleagues, as well as left-wing intellectuals in the West, are impressed by the freedom-crushing potential of ineffectively restrained private power. The dilemma they face is to find a plausible source of restraint. For they are equally sensitive to the tendency of private power holders to colonize the state and thereafter combine its resources with their own. The dilemma, as Dr Chamarik points out, is particularly acute in third-world countries:

Within advanced industrial countries, hegemonic and exploitative relationships have been qualified and restrained somewhat within a democratic framework of civic and political participation. Most of the Third World's developing countries, by contrast, are under authoritarian regimes and traditions, and practically all the public decisions are left to the tiny groups of so-called modernizing elites . . . which looks to foreign capital and its accompanying science and technology as the agent of change and modernization.[8]

Imported science and technology and the modes of thought that accompany them, he goes on to contend, provide the élite with additional means to exploit the great bulk of their fellow citizens. The "modernizing" imports simultaneously undermine the majority's unrealized capacity for improving the material conditions of its life and enhancing its freedom of choice by converting new information about the natural world into appropriate technologies.

We can see, then, that the view exemplified by Dr Chamarik, a view that is very much in the mainstream of third-world discourse about human rights, has two essential elements. One is the proposition that, under prevailing conditions, the development and diffusion of scientific knowledge and the technology stemming from it progressively diminishes the opportunities of most people to impart shape to their lives. The second is the proposition that this morbid consequence of science and technology derives from the capitalist centre's hegemonic relationship to the host of peripheral participants in the global system of political economy.

While the first proposition is not alien to mainstream political discourse in high-technology capitalist states, that discourse is not consistently conducted in the idiom of human rights and, unlike third-world discourse, it ranges over a narrower set of issues. Rather than seeing in science and technology a vast diffuse threat to the possibility of a dignified human existence, opinion leaders in the West worry about very specific threats to a way of life viewed on the whole as historically unexampled in terms of individual autonomy and dignity.

AREAS OF TENSION BETWEEN SCIENCE AND TECHNOLOGY AND HUMAN RIGHTS

In the capitalist democracies, the sense of connection between advances in science and technology, on the one hand, and human rights on the other is clearest in two areas: (1) procreation and child-rearing (what might be called the "biological issues"); and (2) privacy.

Each has a literary antecedent: for the first, Aldous Huxley's *Brave New World*, with its anticipation of genetic selection; for the second, George Orwell's *Nineteen Eighty-four*, with its vision of a state apparatus endowed with the means and the will to conduct perpetual punitive surveillance of its defenceless citizens.

Potential uses of recently acquired knowledge about human biology threaten several of the rights enumerated in the International Covenant on Civil and Political Rights,[9] including the right to life (Article 6), the right not to be subjected "to medical or scientific experimentation" (Article 7), the right to security of the person (Article 9), and the right not to be subjected "to arbitrary or unlawful interference with . . . privacy [or] home" (Article 17). The vast expansion of means for exercising surveillance over individuals and associations (including tracking their movements and intercepting their conversations) and for collating, storing, and rapidly distributing resulting data directly threatens the right to privacy and tends to suppress the exercise of the rights to freedom of speech and association. One must, of course, recall that the same body of theoretical and applied knowledge which poses such risks to the rights enumerated above is also available to enhance the security and welfare of our species.

I propose to use the brief compass of this paper primarily to discuss the biological issues, as well as one other which is increasingly the subject of public debate on both ethical and prudential grounds – the issue of nuclear weapons. Its inclusion may seem paradoxical in the light of my decision not to consider third-generation rights as such. Can nuclear weapons be evaluated within a human rights framework unless one concedes the existence of a right to peace?

The answer, I submit, is yes, for two reasons. One is the erosive effect of preparation for nuclear war on democratic values, particularly the right to meaningful participation in the political process.[10] The nuclear threat aggravates the apparently natural tendency toward the centralization of political power and secrecy about its uses.

A second reason stems from the fact that human rights law is one branch of a larger body of norms whose other branch is the humanitarian law of war.[11] While their development has been roughly coincident over the past century, norms governing wartime treatment of civilians and persons rendered *hors de combat* – i.e. persons deemed innocent and hence deserving protection to the highest degree possible from the dangers incident to armed conflict – acquired legal form far earlier than the norms governing the treatment of persons by their own governments in time of peace.[12] In a world where armed conflict, internal and external, remains almost commonplace and where a very large percentage of research and development is funded by military establishments and then applied at the greatest possible speed, it might seem not merely paradoxical but perverse to discuss the impact of science and technology on human rights without mentioning an impact that could consume through omnicide the subjects of every sort of right.

Even if one puts aside so apocalyptic a view of the nuclear problem, nuclear weapons and strategies can be assessed without any implied affirmation of third-generation rights. Assessment can be conducted entirely within the framework of humanitarian law – well developed and firmly established, animated by the same moral impulse that generated the Human Rights Covenants, an impulse to protect the basic rights of individuals beginning with the rights to life and personal security.

HUMAN RIGHTS AND NUCLEAR WEAPONS

Of course the right to life is not absolute. The state acting through the medium of its police force may take the life of someone violently resisting a proper arrest or attempting to inflict grievous harm on another person. Similarly, in the course of an armed conflict, soldiers authorized by their respective political communities kill each other without violating the right to life.[13] The victims in both contexts are exempted by their act or position (combatant ready and willing to fight) from the category of protected persons. One might say they lose their formal status of "innocents."

Discrimination in war between combatants and non-combatants (civilians and members of the armed forces rendered *hors de combat*) is a central requirement of our moral and legal traditions.[14] Failure to distinguish was among the crimes for which members of the Nazi High Command were punished after the Second World War.[15] Hence the appearance of nuclear weapons, possibly the most portentous product of modern science and technology, immediately posed grave ethical and legal issues.[16] The earliest versions of the weapon had so large a killing range, even in the smallest producible size, that they could not be used with discrimination in densely populated areas. Furthermore, even if they were used in an area occupied almost exclusively by troop or other legitimate targets, the uncontrollable dispersion of radioactive particles menaced the health of civilians far from the zone of conflict, including the inhabitants of neutral countries. The most vulnerable civilians, moreover, were those with the least ability to contribute to war-making activities and hence, in light of the rationale behind the laws of war, the most innocent: the very old and young.

Not only did nuclear weapons challenge the principle of discrimination, in addition they raised difficult questions of justification under two other central principles of normative restraint on the use of force: the principle of proportionality and the principle of no gratuitous injury.

The former requires a reasonable relationship between damage caused and end sought, at the grand strategic no less than the tactical level of action. In other words, states can violate the principle either by using an enormously destructive force – causing incidental and unintended but foreseeable and terrible injury to non-combatants in the vicinity of legitimate targets – to secure some battlefield objective unlikely to affect the overall military balance *and* by using such force to win a war fought over relatively inconsequential political issues.

While the principle of proportionality operates primarily to protect non-combatants, the principle of no gratuitous injury was important to combatants and non-combatants alike. Assuming the equal efficiency of weapons systems or tactics, an army had to employ the one calculated to minimize suffering and permanent injury. One can see, therefore, why a weapon that inflicted horribly painful and permanent injury through burning and radioactivity might be challenged under this principle, although in many or most cases it could be defended successfully by invoking the offsetting principle of relative efficiency.[17]

The first and to date only use of the new weapon was as an instrument of terror designed to deliver a decisive shock to the enemy's political system. Given the stakes in the Pacific War and the efficacy of the attacks on Hiroshima and Nagasaki, the nuclear assault on the inhabitants of those cities seems to have complied with the principle of proportionality, at least in so far as the first bomb was concerned. But of discrimination there was none.[18]

The development of nuclear bombs was carried out in great secrecy. Moreover, right up to the first test, many involved in the project were uncertain whether they would work at all, much less what would be their precise effects.[19] So this extraordinary new weapons system was incorporated into the American arsenal before military and civilian planners were able to think through its possible tactical and strategic roles. As long as the US enjoyed a nuclear monopoly while possessing only a small number of bombs, there was little incentive to see them as anything other than city-busting weapons, and hence, from the perspective of the humanitarian laws of war, as latent delinquents.

Soviet acquisition of nuclear weapons, continuing increase on both sides in the number of nuclear devices, development of missiles as delivery vehicles, and discovery of means to miniaturize the devices, vary their explosive force across a broad spectrum, and sharply reduce their radioactivity created an extraordinary demand for doctrinal development.

Theorists responded rapidly to the new incentives for concentrated thought. Indeed, as Lawrence Freedman demonstrates in his magisterial survey of *The Evolution of Nuclear Strategy*,[20] most of the doctrinal issues which animate contemporary debate were delineated in that efflorescence of theory that marked the decade beginning roughly in the mid-1950s. The most fundamental issues, then as now, were whether nuclear weapons should be used exclusively to deter their use by the Soviet Union, as distinguished from deterring aggression by any means, and whether deterrence for whatever end was most likely to succeed by targeting Soviet cities or Soviet weapons and, in the latter case, by preparing to strike first or by preparing only to retaliate for Soviet first-use.

Neither in explicit doctrine nor force structure has the United States manifested a clear choice among these alternatives. It has, however, varied its inclination. During the McNamara era, the inclination was strongly in the direction of deterring only a nuclear attack and only by means of a retaliatory strike against Soviet population and industrial centres. The acronym MAD – i.e. mutual assured destruction – summarized this inclination.

A powerful force restraining the US from a total commitment to MAD, even under a Secretary of Defense who seemed to doubt that the actual as opposed to the threatened use of nuclear weapons could ever be a rational act of policy, was our commitment to defend Western Europe and the disinclination of our European allies to develop conventional capabilities plainly sufficient to stalemate Soviet conventional forces. Their disinclination was not merely a matter of popular preference for butter over guns. Having experienced the horrors of two wars fought without nuclear weapons and therefore able to envisage the damage a third such war would inflict on their lands where, after all, such a war would be waged, West European élites preferred what they took to be the slight danger of nuclear deterrence failing to prevent conventional attack to what they assumed would be an enhanced risk of conventional war if the threat of recourse to nuclear weapons in the event of a Soviet breakthrough were eschewed.

The felt need to retain a credible first-use option undermined adoption of a force structure designed exclusively for city-busting. For as the Soviet Union's ability to deliver nuclear strikes against the United States expanded, the credibility of our first-use threat was bound to diminish. Charles de Gaulle was only the first European leader actually to say that West Europeans could not expect the United States to sacrifice New York in order to defend Frankfurt. If that trade-off seemed incredible to them, they had to assume that it would be little more credible to Soviet authorities. Thus commitment to first-use in the event of a Soviet conventional thrust into Western Europe was bound to push the US toward a counter-force strategy, since it would provide at least some basis for believing that nuclear weapons could be employed without triggering mutual national suicide. This belief rested on the fact that if either side initiated a nuclear conflict by striking weapons and avoiding major population centres, the latter would function like hostages to tolerable conduct by the recipient of the strike, tolerable conduct being defined as a proportionate response against military targets. To be sure, hostages are sometimes sacrificed; but they would at least heighten the attractions of an invitation to a limited nuclear duel. The Soviet Union did what it could to discourage such a belief by insisting that general nuclear war would inevitably result from any crossing of the nuclear threshold. In any event, some advocates suggested, if for one reason or another a city-banging exchange began, counterforce-capable weapons and targeting might allow some reduction in the level of damage sustained by the United States.

Even as a deterrent to a first strike against the United States, however, a city-busting strategy had a credibility problem. If deterrence failed and a Soviet first-strike shattered the United States, what rational purpose would be served by exterminating the people of Russia? One could, of course, transcend the problem by following the *Dr Strangelove* scenario of eliminating the human factor and automating the response. But after having warning systems triggered by errant geese, this option could not have much appeal even in the rarified atmosphere of nuclear war theorists. To most people, automaticity seemed an unnecessary as well as an unbearable risk. Vengeance may not be a "rational" purpose, but it certainly is a common motive, sufficiently common to deter rational opponents.

By the late 1960s, the elements of a strategic equilibrium were in place. Missiles formed the core of each side's nuclear force. They were protected from a disarming first strike, in part by deployment in hardened silos on land and on submarines capable of using wide stretches of the globe's oceans for concealment, but in larger part by the roughly equivalent number of warheads and delivery vehicles on both sides. That numerical equivalence, coupled with certainty that a substantial percentage of missiles would abort on firing or would go astray after firing, meant that a disarming strike would more effectively disarm the initiator than his intended victim. And since neither side could defend itself against missile attack, and revenge, together with a powerfully programmed commitment to retaliation, made a devastating response to any first strike psychologically plausible, the risks associated with a first strike appeared wildly disproportional to any conceivable gain.

Both sides had multiple targeting options. They could hit weapons or cities. But the former option, if it made sense at all, did so only as part of an escalation scenario initiated by a conventional Soviet assault in Europe that could not be repelled by conventional means or tactical nuclear weapons. In other words, as I suggested above, a launch against a limited number of weapons sites located far from population centres might halt a Soviet attack by demonstrating resolve without inducing fear that the West was using the occasion of a conventional war to win a nuclear war.

Although escalation scenarios were built into NATO's strategic doctrine,

Mutual Assured Destruction was in fact the dominant theme of nuclear policy on both sides. The anti-ballistic missile arms control treaty of 1972 seemed to institutionalize the ascendancy of its supporters over advocates of a nuclear-war-fighting capability. Eschewing defensive measures, both sides seemed to accept the logic of MAD and its concomitant moral dilemma, the deliberate targeting and declared intention to annihilate civil society in response to a first strike.

The widespread belief that, under then prevailing technological conditions, MAD had sharply reduced the risk of nuclear war did not still all dissident voices. Certain theological moralists (the Protestant Paul Ramsey[21] was prominent among them in the 1970s) declared unacceptable any system resting on the threat to attact civilian populations. The moral tradition for which they spoke condemned the intention to do evil no less than the doing of it. But indictment of MAD led to conflicting convictions: on one side, that nuclear weapons had to be banned; on the other, that the West had to develop weapons and strategies which would permit it to fight defensive nuclear wars without violating the traditional principles of the law of war.

Advocates of the second position conceded that, as in any major war, attacks on military targets were, under the best of circumstances, bound to cause collateral injury to the civilian population. But under the principle of the "double effect," such injury was morally tolerable because the foreseeable incidental effect on non-combatants was unintended. And, the argument continued, so long as belligerents complied as well with the principle of proportionality, the distinction between moral and immoral use of force could in practice be maintained.

Advocates of prohibition, on the other hand, believed that, once the nuclear threshold was breached, even at the tactical level, there was an intolerable risk of escalation to full-scale strategic nuclear war. In such a war, the foreseeable damage to civil society within the belligerent states and probably to neutral states as well was so vast that the conflict could not possibly comply with any reasonable conception of moral fit between means and ends. Discrimination between combatants and non-combatants was not only a virtuous end; in addition, it bonded with the principle of proportionality to bar wars of societal extermination. Thus, however unintended the damage to non-military targets might be in a nuclear attack, the attack could not help violating the spirit if not the form of the discrimination principle.[22]

Support for a nuclear-war-fighting doctrine and force structure did not stem exclusively from theological convictions. Since MAD is a strategy calculated to achieve virtually a de facto ban on superpower use of strategic nuclear weapons – achieved, of course, through the medium of continuing possession – it was bound to cause unease among many officers on both sides. At least in great powers, military training has always had an offensive bent. While the ultimate political ends of war could as easily be defensive as aggressive, the military end was victory. To achieve victory one procured the maximum firepower available and deployed it in the way best calculated to break the will or physically to destroy the opponent's armed forces. At the same time, one sought to minimize the damage that the other side could inflict either on your own forces or its civilian base. Procuring offensive and defensive systems compatible with strategic nuclear initiatives is inconsistent with the rationale of MAD. So, it could be argued, is a foreign policy designed to threaten the fundamental interests either of Soviet or American élites.

MAD, its defenders believe, makes the prospect of nuclear war so awful that rational leaders will approach with extreme caution policy options that could hurl the superpowers onto a rising and possibly uncontrollable escalator of violence. But intense stress induced by a sense of confronting a crisis diminishes the capacity for rational calculation. Persons in the grip of crisis psychology are unusually liable to misconstrue the acts of others and to misperceive the implications of their own. Diminished rationality is, therefore, one risk MAD cannot abolish.

A second is desperation induced by the accurate belief of one superpower regime that the other is threatening its political survival. At that point, rationality becomes problematic. Confronted with great risks, reasonable people may rightly conclude that the only way to avert one disaster is to threaten (including steps to initiate) measures that could lead to another.

It is, therefore, not surprising that the division between supporters of MAD and supporters of a nuclear-war-fighting strategy seems to coincide roughly with the division between supporters and opponents of US–Soviet détente (by that name or another). The latter, apparently convinced that the Soviet regime is driven by its very nature to hammer unrelentingly on the vital interests of the capitalist democracies,[23] assume that the safest course is ceaseless counterpressure. MAD, they believed, tended to induce in Western polities a sanguine passivity. That passivity not only made the democracies vulnerable to conventional threats and subversion but, *a fortiori*, discouraged adoption of positive measures to reduce Soviet capacity to mount threats. And MAD reinforced optimism with the fear that driving the Soviets into a political corner would induce the very high-risk behaviour MAD was designed to avert.

The Technological Erosion of MAD

The treaty to ban deployment of anti-ballistic missiles seemed to stabilize the strategic dimension of superpower relations. The Moscow joint declaration of a *modus vivendi* in the competition for global influence seemed to stabilize their political dimension.[24] MAD thus appeared as a central feature of détente. But at the very moment when the superpowers seemed to be entering a new, comparatively tranquil phase in their relationship, the inertial force of technology and the unresolved contradiction of politics were laying the foundations of a new Cold War.

The political issues fall outside the compass of this paper. (I have touched on them in earlier works and will explore them more fully in a book I am now writing.) I will address only the technological ones.

The explosion of public controversy over the Reagan Administration's "Star

Wars" initiative,[25] which, after all, is still in an early developmental phase, contrasts oddly with the virtual silence that accompanied actual deployment of multiple, independently targeted re-entry vehicles (MIRVs). For they immediately and radically altered the numerical relationship between first- and second-strike strategies from which MAD had evolved. Perhaps it is the retrospective appreciation of just how grave a development MIRVs represented that has powered the reaction to Star Wars; for the "Strategic Defense Initiative" (as it has been labelled by the Reagan Administration) threatens to complete what MIRVing began, namely the elimination of MAD as the doctrinal instrument for keeping the nuclear peace.

As I indicated above, assuming rough numerical equality of delivery vehicles and warheads, before MIRV an all-out first strike against the opponent's strategic forces would be self-disarming. MIRVs plainly changed that arithmetic. If, for instance, each side had 1,000 ground-based missiles at known locations and each missile has five independently targetable warheads, either country could launch a first strike using only 40 per cent of its strategic force and still allocate two warheads to each of its counterpart's missiles. The conspicuous increase in accuracy both sides were achieving in test firings of individual missiles made it appear likely, at least on paper, that, by virtue of duplicate targeting, such a strike would destroy all or nearly all of its object's land-based strategic forces, the most powerful and accurate element in its strategic arsenal.

So much for hypothetical cases. In the real world, where US strategic forces are a triad of land-based missiles, submarine-based missiles, and bombers, the US could absorb a Soviet first strike that destroyed all of its land-based missiles and all bombers not in the air or on ready alert and still be left with thousands of sub-launched warheads, enough to convert the Soviet Union into radioactive wasteland. (As demonstrated by the immense dispersion of radioactive materials released by the nuclear accident in Chernobyl, the nuclear exchange would also have horrendous consequences for many other countries, if not the entire earth.) Since MIRV did not affect either side's capacity to deliver a fatal retaliatory blow, why should it have undermined the conviction of stability induced by MAD? Doubt stemmed from the following theory: A Soviet first strike employing only a portion of its land-based missile force could wipe out its US counterpart while leaving the main American urban centres intact. To be sure, collateral damage from the anti-weapons strike would have left millions of American dead and dying; but, it has been argued, the bulk of the population would be intact and thus hostage to the threat of Soviet counter-city retaliation if the President should order the launch of sea-based missiles against Soviet urban centres. An attack against remaining Soviet missiles would be only partially effective, SDI proponents claim, because sea-based missiles are neither as powerful nor as accurate as those based on land.[26] Submarines, moreover, have no at sea reload capacity. So even if the President ordered counter-weapons retaliation, the net result would be a force ratio advantageous to the Soviet Union. For it would still deploy part of its land-based force and all of its own sea-based strategic weapons.

"So what?" some experts asked. How could this bean-counter's triumph be converted into the currency of political gain? The only targets left in the US would be its cities. At this point, at least, the logic of MAD would take over. The Soviet Union would be no better positioned than before to coerce the United States by threatening those cities, since the United States would have retained its capacity to retaliate. MIRVs should not be seen, therefore, to have affected the stability of strategic nuclear relations.

To this scepticism about the supposed consequences of MIRVs, there were several possible responses. One was little more than the naked assertion that, rightly or wrongly, most people are impressed by numerical differences. Hence, in the hypothesized case, most people would perceive a Soviet victory. That perception would shatter confidence in US security guarantees. NATO would unravel and old allies of the US would seek accommodation with the Soviet Union. It was further argued that the scenario itself would undermine West Europeans' confidence in the US commitment to their defence, relying as it does, theoretically, on readiness to employ strategic nuclear weapons against the Soviet Union.

Alternatively, it would frighteningly heighten the risk of unintended nuclear war by leading to adoption of a fire-on-warning posture. Another way of countering the first-strike risk was to develop mobile land-based missiles which could also be hidden. That, however, would complicate arms control verification and would sharply increase the tendency toward greater secrecy in government, with its associated consequences of closer screening and surveillance of government employees and the employees of government contractors. In short, because MIRV mattered in the minds of military experts, it mattered for all people, at least in the superpowers and probably in the entire world.

Whether the anxieties induced by the MIRVing of delivery vehicles in the first half of the 1970s contributed in any significant degree to the atmospherics that surrounded the crack-up of détente in the second half is problematic. What one can say with some confidence is that the crack-up, culminating in the Soviet invasion of Afghanistan, heightened those anxieties and thereby eased the way on to centre stage of hitherto marginalized advocates of strategic defence. Whether in the longer term the genie of strategic defence could have been contained in its bottle is subject to doubt. Now that it is out, the question is whether it is likely to reduce or enlarge the human rights threats generated by nuclear weapons.

Research and development, doctrinal debate, and superpower negotiations are at too early a stage to allow any self-assured answer to that question which does not smack of hubris. What one can do is identify the issues that lie at the heart of the matter. One, certainly, is the problem of transition from MAD. In developing and deploying defensive systems, the superpowers will not move in unison unless they agree to such an arrangement. In the absence of agreement, somewhere along the temporal continuum of research, development or deployment, one of them may well achieve what appears to be a serious advantage.[27]

Only President Reagan and a few other ecstatics appear to envisage anything like a secure umbrella.[28] What SDR advocates do claim is that the technology developable on the basis of extant theory allows the United States to deploy, well before the end of the century, a defensive system that would offset the advantages conferred on the offence by MIRVs. Such a system could, moreover, deal with single missiles launched by accident, by mavericks, or by third states.

The capacity to moderate the threat of accidental or unauthorized missile firings by either superpower – as well as covert launchings by terrorist regimes, whether for purposes of blackmail, revenge, or to catalyse a superpower conflict – must be deemed a contribution to human rights. But a transition marked by unilateral, perishable breakthroughs in research, development, or deployment seems certain to heighten the risk of nuclear war. Paradoxically, the partial character of any foreseeable defensive system could prove more deleterious to stability than an airtight system, assuming the latter's capabilities were appreciated only after its deployment.[29]

The problem with a partially effective system is its Janus-like quality. On the one hand, by reducing the advantages of a counter-force first strike, it promises to reduce anxiety and the temptations of such dangerous expedients as launch on warning. On the other hand, precisely because it will be most effective in countering a nuclear force decimated and disorganized by a first strike, it promises to increase anxiety and, in the event of a crisis, to add in some immeasurable degree to the temptations of a first strike.

In a 1987 article, Robert Gromoll, an astute participant in the ongoing debate, argued that the "dynamics of strategic uncertainty" are still more complicated than even my last remarks suggest.[30] While the status quo presents each side "with the possibility of only three basic strategic relationships – offensive equality (or parity), offensive superiority, and offensive inferiority" – once each side begins deploying ballistic missile defences, any one of nine strategic relationships will obtain at a given time. And of that nine, only three "can be deemed both truly stable and acceptable" to both parties in terms of defence of its most fundamental interests.[31] Furthermore, since both sides will labour assiduously to improve their systems, transitions from one relationship to another will recur.

"The many uncertainties that (ballistic missile defence) would spawn," he fears, "could foster irrational national postures founded on boastful overconfidence, wishful thinking, and unhealthy pessimism just about as easily as they could foster more prudential assessments of strategic realities."

In conjunction with the many ambiguities surrounding [ballistic missile defence's] likely performance in actual combat, these political and psychological variables raise greater potential for misperception of relative offensive–defensive capabilities and miscalculation of risks. Soviet and American leaders could easily arrive at notably different *perceptually wrought* conclusions about what [ballistic missile defence] could do, and then proceed on the basis of asymmetrical and dangerously inaccurate strategic assessments.[32]

For the foreseeable future, only the US and the USSR will have the capacity to deploy ABM systems. In the eyes of reasonable people, the Janus-like quality of such systems adds an important incentive for the superpowers to emphasize far more strongly the co-operative rather than the competitive aspects of their relationship. As the two principal beneficiaries of the global system that arose from the ashes of the Second World War, they have a powerful vested interest in co-operating to contain the anarchic forces threatening that minimum world order without which the prospect for human rights will be bleak. We cherish the hope that self-interest will prove stronger than ideology.

THE BIOLOGICAL ISSUES: MAKING LIFE, SUSTAINING LIFE, AND REDEFINING THE FAMILY[33]

Any orderly discussion of the human rights issues raised by developments in science and technology affecting conception and gestation should proceed in three phases: identification of the relevant developments; description of their present and potential uses and the social purposes of such uses; evaluation from a human rights perspective.

The key developments concerning fertilization can be summarized as the capacity to collect and preserve sperm, to collect and preserve ova, and to unite them in the body of choice or outside any body (fertilization *in vitro*, the so-called "test-tube baby"). The key developments concerning gestation are the capacity to move the fertilized ovum to a body of choice, to remove the foetus from the womb months before term and sustain it indefinitely, to identify many important characteristics of the foetus (for example, sex and genetic anomalies), and to intervene during gestation in order to reduce or eliminate those anomalies prior to birth.

A central objective of these developments has been to provide couples, one or both of whom are infertile, with the opportunity to acquire children and thereby realize the traditional conception of full family life. Despite the great worldwide pool of children available for adoption, many couples apparently experience an intense preference for a child who will carry the genetic endowment of at least one parent.

The choice of means for satisfying their preference is, at least in part, a function of the cause of their inability to conceive. If, for instance, it results from some sort of blockage that impedes normal ovulation, doctors can extract ova, fertilize one with sperm from the husband and replace the fertilized ovum in the wife's womb for normal gestation.

The Vatican's Congregation for the Doctrine of the Faith has condemned this measure along with all other "artificial" forms of procreation as threats to the sanctity of the conjugal act, to the family, and to human dignity. I will discuss its position below together with other views on what one author has called "the ethics of human manufacture."[34] At this point I will simply note that from a strictly utilitarian perspective, in vitro fertilization raises two concerns. One is whether children so conceived have an above average risk of congenital anomalies. I gather that in theory there is no reason why this should be so. A second concern is the potential for foetal experimentation arising from the possibility of fertilizing more than one ovum. The extra fertilized ova, the embryos, could then be frozen and in that state preserved for experimental purposes. Both governmental and private groups concerned with the ethical issues arising from the new capacity for human manufacture have recognized the danger of what they uniformly regard as intolerable abuse and have proposed regulations I will consider shortly.

Where either parent is sterile, but the wife can gestate, conception obviously requires the services of a third party. But that connection is so brief and at such a preliminary stage as to minimize the third party's psychological involvement with the issue of this process. And, although the resulting child's legal and biological father or mother will not coincide, he or she need not know that, since they will experience a unity of gestation, birth, and rearing.

The risk of psychological trauma for one or more of the parties (including the child) involved in the process of conception, gestation, and parturition sharply increases when the wife is unable to gestate or where gestation and/or parturition would be physically dangerous or where, for subjective reasons of one sort or another, she does not wish to undergo the child-bearing experience. Then the embryo must develop within the body of another woman (often referred to as the "surrogate mother") who agrees before the fact to surrender the child following parturition. Limited experience to date suggests what one would in any event have assumed, that giving up the child may prove far more difficult emotionally than the surrogate had anticipated. Legislators, judges, and scholars in the United States are now struggling to decide how the resulting conflicts should be resolved.

As the variety of reasons for use of surrogate mothers illustrates, the new biological techniques can be applied to ends other than overcoming natural obstacles to reproduction within the traditional family.

One is permitting a single person to become the parent of a genetically related child even if that person cannot find a sexual partner or does not wish to conceive through sexual intercourse. Among the many consequences of these techniques, it would appear, is circumvention of existing regulations governing adoption and, in particular, inhibiting adoption by single parents.

A second unconventional end is permitting unisexual couples to form bi-generational, genetically related family units. Public efforts to block this contingency will sharply focus the claim that any government action implying discrimination on the basis of sexual preference violates human rights of association, family formation, and privacy.

A third unconventional end is facilitating genetic and sexual selection by parents or, in imaginable totalitarian political and social settings, by external agencies. As I indicated above, the sex, the genetic, and other characteristics of the foetus can now be determined in the early stages of gestation. Like every expansion of scientific knowledge and technique, this one is available both for benign and for morally dubious applications. To the extent the new knowledge permits correction during gestation of foetal anomalies that would otherwise cause early death or mental or physical crippling, it is in itself benign, although the costs of intervention and of the infrastructure that will make it possible are bound to aggravate extant inequalities in the availability of medical assistance. To the extent that knowledge leads to sexually discriminate abortion in favour of males, as many fear it will, the discrimination itself and its unpredictable effects on the social order are sources of profound concern. Even more ominous is the potential use of abortion as part of a total programme of genetic selection directed by the state or some informal centre of authority (for example, the leaders of a religious sect).

A fourth unconventional end of the new knowledge and techniques is the production and preservation of foetuses for medical experimentation and as a source of tissue for transplants, initially in the treatment of Parkinson's disease and other nerve disorders. But the potential use goes much beyond that. As the *New York Times* reported on 16 August 1987, current research involves the use of foetal brain, pancreas, and liver tissue to treat Parkinson's disease, Alzheimer's disease, Huntington's chorea, spinal chord injuries, diabetes, leukaemia, aplastic anaemia, and radiation sickness. According to Tamar Lewin, scientists "expect foetal tissue to be particularly valuable in implant treatments because it grows faster than adult tissue, is more adaptable and causes less immunological rejection."[35]

This potential for malign no less than benign applications generates crescendoing political conflict that will increasingly express itself in legislative fora. Consensus seems to be forming only with respect to the issue of foetal manufacture through *in vitro* fertilization for the specific purpose of producing a reservoir of tissue for transplantation. Most secular ethics committees in the English-speaking world that have considered the issue have recommended that up to 14 days after conception, the embryo be considered tissue that may ethically be disposed of or subjected to research; they have also recommended that experimentation after that time be made a criminal offence. The Warnock Committee in the United Kingdom, an ethics board set up by the US Department of Health, Education and Welfare, the 1984 Victoria (Australia) Committee, the 1985 Ontario Law Reform Commission, and the American Fertility Society in 1986 all favoured the 14-day rule. France's National Ethics Committee recommended seven days.

The perceived virtues of the rule are the fact that it provides a bright line and the line is not arbitrary. As the American commentator Charles Krauthammer points out in a provocative article on "the ethics of human manufacture":

Fourteen days marks the development of the "primitive streak" after which twinning is no longer possible and neural development begins. It is also important clinically. The technique of *in vitro* fertilization involves taking an ovum and sperm and mixing them in a laboratory. Because the "take" upon implantation in the uterus is difficult, more than one ovum is fertilized. The clinician watches the dish in which the process of fertilization occurs to see which pre-embryos (we might call them) are developing best and implants them. Without the 14-day bright line, discarding these "spare embryos" would have to be considered murder.[36]

Of course the last sentence is a non-sequitur. Logic does not require that we recognize only the polar positions of unrestricted right to use embryos and treating the embryo when it is still only a tiny mass of cells as a person subject to being murdered. Krauthammer's second argument – "clinical importance" – calls for some bright-line rule but not any particular one. But that still leaves us with at least one objective reason for the 14-day rule. Incorporation of that rule into law would, however, create at least one striking anomaly if women simultaneously enjoy an unlimited discretion to abort at least during the first trimester following conception. For in that case, a woman could legally choose to generate foetal tissues by conceiving, whether through artificial insemination or otherwise, and then aborting on the ninetieth day. A clinician, on the other hand, acting at the request and with the co-operation of the woman providing ova for in *vitro* fertilization, would be barred from achieving the same end unless he could act within 14 days.

Although vivid, the anomaly can be justified. Any attempt to administer a single rule in both cases would produce an appalling invasion of a family's privacy and would still not be capable of anything but the most arbitrary application. If a woman may abort for no stated reason at all, but is prohibited from aborting for the "wrong" reason (in my hypothetical case the reason being to generate tissue to aid in treating herself or someone else), a law applying equally to mother and clinician obviously would require the public authorities to establish the former's motive in seeking abortion. In attempting to protect the human rights of the foetus, the state would engage in massive violation of the right to personal and family privacy.

In addition, one might justify the anomaly on the grounds that the probability of foetal abuse is far greater where fertilization is achieved in *vitro* and the resulting embryo develops outside the womb. Ethical and emotional factors, as well as the physical burden of gestation, are certain to constrain tissue production through abortion. However, among the underclass in affluent states and among the general population in poor states, the terrible pressure of poverty may erode those constraints.

If tissue implantation proves to be a breakthrough in the treatment of the various illnesses enumerated above and if foetal tissue proves to be far superior to any feasible alternative, demand could soar at an extraordinary rate. And if affluent states prohibit tissue production through in *vitro* fertilization (or go further still and prohibit the use of tissue from an aborted foetus), economic pressure might drive impoverished women to serve as human incubators from whom embryos will be harvested. This could occur in conjunction with the establishment of clinics for *in vitro* fertilization in countries with permissive legislation. Some evidence now exists that the most useful tissue will come from foetuses that have developed beyond the first trimester. Entrepreneurs operating in the poorest countries might decide it was more efficient to use human rather than artificial wombs.

Laws banning the importation of foetal tissues are unlikely by themselves to combat this phenomenon effectively. If the contingencies enumerated above come to pass, *and if present efforts to reproduce artificially cells from foetal tissue are not successful,* the intensity of demand for foetal tissue will rival and could easily surpass the demand for cocaine. The world will then be faced with another form of unstoppable contraband, another gigantic global black market.

As I indicated in my introduction, at least four of the substantive provisions of the International Covenant on Civil and Political Rights – the right to life, the right not to be subjected to medical or scientific experimentation, the right to security of the person, and the right not to be subjected to arbitrary or unlawful interference with personal and familial privacy – bear in varying degrees on issues arising from the rapid advance in knowledge and technology affecting the process of life creation. But, as I also suggested, more is at stake than those four rights or, indeed, all the rights codified in the Covenant and other human rights instruments. The very idea of human rights arguably rests on a certain conception of what it means to be human, on a reverence for life and for autonomy, all of which could conceivably be undermined if the production, maintenance, manipulation, and termination of embryos became commonplace, and particularly if this activity were carried on to a considerable extent for profit. Parallel expansion of the extant semi-clandestine market in body parts will aggravate that risk. One way of putting the issue is whether the idea of human rights will continue to enjoy powerful support in an environment characterized by the commodification of the body.

In addressing these issues, the Vatican, through a document from its Congregation for the Doctrine of the Faith, has argued that any interference whatsoever with what it regards as natural forms of procreation will pitch humanity onto a slippery slope down which it will then tumble into a condition of moral numbness; by disregarding the Kantian principle that no human being can properly be used as a means for another's ends, ever more heinous acts will be committed in the name of utility. As a guide to all humanity rather than simply its flock, the Church's position is besieged by difficulties. Only one of them is the refusal of most people, including many Catholics, to treat the few clustered cells that are the immediate result of conception as morally indistinguishable from a fully formed foetus in its third trimester, a new-born infant, or a person in the prime of life. The view that life begins absolutely at conception evokes passionate but minority support.

A second is that the slippery-slope argument, almost wherever it is applied, is not really an argument but a raw claim so broad and diffuse as to defy verification. The claim's breadth is so full of potential to stultify human ingenuity, so indifferent to other moral claims in particular cases, and therefore so damaging to tangible interests that it invites rejection.

Krauthammer argues plausibly that "the Vatican document . . . is a radical act of resistance to the technological hubris of modern reproductive medicine . . . The Church opposes everything: *in vitro* fertilization, embryo freezing, embryo transfer, surrogate motherhood, artificial insemination not just by donor but even by husband."[37] It is no less opposed to artificial aids in creating families than to their use in preventing conception. Yet both are widely regarded as means for contributing, on the one hand to a fuller and richer existence, and on the other to the avoidance of cruelty and suffering.

While himself rejecting slippery-slope reasoning in some cases, Krauthammer finds the argument decisive in those where it is proposed to hasten imminent death or end a vegetative subsistence for one person in order to effect a life-saving transfer of their vital organs to another.[38] The issue came into public focus when legislators in two American states introduced legislation that would allow removal of vital organs from anencephalic babies (children born with so little brain tissue that, even if they survive birth, they cannot live longer than a few months – 95 per cent die within one week) before brain death is certified. If doctors wait until "whole brain death" has occurred, the organs will not be usable. Whether Krauthammer would arrive at the same result where the life of an anencephalic child is temporarily maintained by aggressive medical intervention in order to preserve organs pending a transplant is less clear. A professional ethicist, Professor James W. Walters, argues that the cases are distinguishable and that the latter should be deemed acceptable as long as the period of maintenance is fixed and, presumably, brief.[39]

The danger of utilitarian apologies ratcheting society into a state of generalized indifference to individual moral claims attends all the new developments in the biological field. Against it, however, stands the promise of incalculable positive contributions to individual fulfilment, as well as broad societal interests. It therefore seems likely that most polities will attempt to isolate the risk without rejecting the entire package in which it is carried. The human species' Promethian passion to better its condition through the systematic application of reason and imagination to the natural world cannot be stilled simply by the invocation of danger. If history demonstrates anything, it demonstrates that. Given the awesome implications for all peoples of the developments under review, the United Nations Human Rights Commission should assist the required effort to find morally compelling points of demarcation. In doing so, it must take account not only of the violations of human rights that could occur as a result of the new and rapidly developing technology, but also of the violations that might occur as a consequence of various possible forms of regulation by the state. It is also necessary to consider the effect of regulation on full realization of the technology's positive potential. Balanced consideration is likely to lead to distinctions between behaviour that should be inhibited through moral censure alone and behaviour that should be regulated, penalized or flatly outlawed by the state.

NEW TECHNOLOGIES FOR THE COLLECTION AND DISTRIBUTION
OF INFORMATION

Writing not long after the Second World War, George Orwell anticipated the extraordinary increase in capacity for the collection, retention, production, and distribution of information and dramatized its dark side. If we take *Nineteen Eighty-four* as a full statement of his views, we would have to conclude that he saw only the evil potential of technologies that also have a vast capacity for enhancing human freedom and security. Any list of the momentous developments already realized or in train would include the following:

1. The accumulation of enormous computer-generated data bases that can be made generally accessible at moderate cost to persons in every part of the globe.
2. The capacity to reproduce documents with perfect accuracy at very low cost on equipment that requires no more space than a desk top.
3. The capacity – resulting from the development of mini-computers, sophisticated software and laser printers that can fit into a single room, coupled with the accessibility of data – to publish newspapers or magazines with high-quality print and sophisticated graphics at a cost which makes it possible for virtually any member of the middle classes in affluent countries to enter the field, albeit, with rare exceptions, only in a very restricted geographic area (where intimate knowledge and locally attractive eccentricity of view can compete against public- and private-sector giants).
4. The capacity to deliver television programmes from a source located anywhere on the globe to receivers in any other part of the globe.
5. The capacity covertly to record telephone conversations, and face-to-face conversations both in public and private places.
6. The capacity for covert surveillance at the micro level of individuals on foot or in cars or of activities of every sort being carried out all over a targeted country or countries. The related capacity to map a country's resource potential and to monitor changes in ecology.

Whether, on balance, these developments will enhance human welfare and widen the realm of individual freedom is problematical. That many governments, including to some degree democratic ones, see serious risks to public order and societal health from their eroding control over the means of communication, particularly the mass media, is evident. One example among many: on 21 September 1987, Prime Minister Margaret Thatcher of the United Kingdom presided over a five-hour meeting of ministers and television executives to discuss what British viewers should be allowed to see when the number of available channels expands from the present four to the anticipated 60 in 1995. One of the participants said after the meeting that: "The government wants as much competition as possible against a background of maintaining standards."[40] "Translated into American," one American commentator subsequently wrote, "that means: 'We accept the inevitability of uncontrolled satellite and cable broadcasting, but we will do whatever we can to stop the game shows, the fictional history of docudramas, the toy advertising, the violence, the pornography that are allowed in the United States.'"[41]

As some governments in the West worry about the loss of control, many private citizens worry about an excess of control, first of information and then of people as a result of the government's increased means for monitoring and recording their behaviour. The tug of war is particularly evident in the United States where, unlike most of Western Europe, there is a powerful tradition at the national and, even more, at the state level of openness in government. The public is perceived as having the right not only to know what decisions are reached but to be privy to the decision-making process. One manifestation of the conflicting values and pressures is the Reagan Administration's efforts to limit access even to unclassified data bases and to narrow the reach of the Freedom of Information Act. Since its adoption that Act had allowed private citizens and organizations to discover that they had been under surveillance by security agencies at one point or another, agencies acting without judicial authorization.

At stake here, of course, is the value broadly characterized as "privacy," a value that occupies one of the central places in that conception of fundamental human rights native to the capitalist democracies of the West. One threat to it is the ease with which governments may dip into private-sector data banks. In the course of a recent public television series commemorating the Bicentennial of the US Constitution, Mr Ira Glasser, Executive Director of the American Civil Liberties Union, summed up the problem as follows:

In the eighteenth century, everybody's personal papers were in their house or in their place of business. Well, most of our personal papers are not at our houses anymore or our places of business. They're on computer disks. They're in the custody of third parties – banks, insurance companies, medical insurance, your employer.

If you want to find out something about somebody, just look at their cancelled checks and their credit card receipts. Now if you kept those at home the government couldn't search. But because someone else is keeping them, the government can go and get them merely by subpoena. You don't know about it and they don't need a warrant. The Supreme Court has been asked to extend the Fourth Amendment protections (prohibiting search and seizure without judicial warrant) to third party custodians of your personal papers, but it has declined to do so on the grounds that the Fourth Amendment only protects your house and your place of business.

So now we have a Fourth Amendment which continues to protect the places where the information used to be kept, but the information has flown the coop. It isn't there any more. All that information is out there floating. And anybody can plug into it. It was one thing when all that information was fragmentary and it got lost. But now the information is persistent. It persists over time and it persists over space. It doesn't go away whether it's accurate or not, it doesn't go away whether it's relevant or not. And it follows you everywhere. And even if it's on a lot of different computers, if you could link those computers up with something like a social security number, all of a sudden overnight you have a national dossier.[42]

Even if the government does not have access, even if the data remains in private hands, given the extent to which each individual is now dependent for his or her survival and ability to serve as a functioning member of society on huge privately controlled enterprises, the new technology's capacity to diminish human freedom is impressive. Some of the principal risks to the individual's capacity to defend against arbitrary deprivation of very important interests by private enterprises are elaborated in a 1987 article by Spiros Simitis, a German professor of civil and labour law and the Data Protection Commissioner of the State of Hesse in the Federal Republic.[43] He begins by stating that "the boundary between a permissible exchange of facts about people, necessary to avoid misrepresentation, and an impermissible intrusion and surveillance is entirely unclear."[44] He then proceeds to offer examples of information–gathering with ambivalent implications.

One example is that of "the transparent patient." It arises out of each country's efforts to contain the exploding costs of health-care delivery. Increasingly, individuals and the services provided to them are measured against computer-designed models. In the Federal Republic of Germany, in cases where the number or type or cost of services exceeds the model, the patients' deviations from the optimum are brought to their attention and they are asked to discuss with a doctor named by the insurance company ways and means of reducing future costs. Also in the FRG, the social security agencies have developed the profile of an ideal cost-saving patient that has led in turn to a form that must be presented to the doctor at the beginning of each consultation. "The form explicitly requires that the doctor confine her services to 'strictly necessary' treatments and enumerates therapeutic measures as well as certain medication that cannot be prescribed without prior approval of the agency."[45] It follows that intimate details of treatment must be shared with civil servants not bound to confidentiality by the Hippocratic oath and the deeply rooted tradition and intense peer pressure associated with it. But the more comprehensive problem is this, that

Such data use results in an entirely transparent patient who becomes the object of a policy that deliberately employs all available information on her habits and activities in order to adapt her to insurers' expectations. The patient is seen and treated as the sum of constantly increasing, strictly formalized, and carefully recorded data that can, at any moment, be combined and compared according to criteria fixed by insurers (whether public or private). Hence, as automated processing is perfected, the patient's position is increasingly determined by a computer-made and insurer-approved, secondhand identity.[46]

Similar developments can be seen in the public sector. Professor Simitis cites a research programme, designed by the Health Council of Oslo, the purpose of which is to collect data about young children that will enable the authorities to identify those children with psychological problems or syndromes that could lead to "anti-social behaviour" as they grow older.[47] He also cites in this connection the recent attempt by the city of Bremen, Federal Republic of Germany, to establish an automatic file of children exhibiting obviously "odd behaviour."

Thefts, excessive aggressiveness, or repeated lying were the identifying signs. Once again, the aim was therapeutic: To treat and to inhibit "dangerous" behaviour and thus to guide the child to adapt better to societal expectations.[48]

In addition to having their privacy invaded, individuals are increasingly threatened with loss of material benefits and of the presumption of innocence in criminal proceedings, all as a consequence of deviation from computer-generated expectations of proper behaviour. Sweden now polices its tax and social security programmes by using automated processing to define "suspect populations" and to trigger follow-ups. To illustrate the consequences, Simitis recalls a "'fishing expedition' aimed at detecting fraudulent housing aid recipients . . ." As the result of matching information from two different sets of records, the government suspected some one thousand persons of having committed fraud. Prosecutions ensued and some persons were quickly convicted. "Ultimately, however, the government had to admit that only one person out of the one thousand suspects was really guilty."[49]

In the American state of Massachusetts, the Medicaid benefits of an elderly woman living in a nursing home were terminated "because, according to a computer match of welfare rolls and bank accounts, her account exceeded the Medicaid asset limit. The account contained, however, a certificate of deposit in trust that was intended to cover her funeral expenses. The computer failed to recognize that under federal regulations the certificate of deposit was an exempt resource not to be included in the calculation of her assets for purposes of Medicaid."[50]

For the billions of desperate people scratching a bare living in the slums and parched fields of the third world, some of the human rights issues generated in the first world by the inexorable advance of science and its technological offspring may seem not only remote but grotesquely trivial. Yet can it not be said that the claims to privacy of comfortable middle-class people in affluent societies, no less than the claims to food and land and shelter of impoverished billions in third-world states, derive their power from a shared respect for each individual as a unique and marvellous being, a moral equal of every other person on earth? We are all partners in the tragic dance to the music of time. To disparage anyone's claim to be a bearer of rights commanding respect and protection is to deny that universal fraternity without which all the covenants and conventions and declarations of the modern human rights movement would be like shallow marks in the sands of a wind-whipped globe.

NOTES

1. Utilitarianism is the most common expression of a still broader moral theory, *consequentialism.* "A consequentialist theory determines the moral status of an action exclusively in terms of the value or disvalue of the action's consequences. An action is morally correct if and only if it brings about a balance of value over disvalue greater

than that of any other action (including doing nothing) that the agent could have performed. Different consequentialist theories characterize value in different ways, with the most popular form, utilitarianism, identifying it with pleasure, happiness, or preference realization. On any consequentialist theory, the use of violence brings about immediate disvalue in the harm and suffering it directly causes. The main condition of limitation on military action set by a consequentialist theory is, then, that the action is permitted only if its short-term and long-term consequences are of sufficient value to counterbalance the disvalue of the violence it directly involves." Avner Cohen and Steven Lee, "The Nuclear Predicament," in A. Cohen and S. Lee, eds., *Nuclear Weapons and the Future of Humanity* (Rowman & Allanheld, Totowa, 1986), p. 1. I will hereafter cite the book as *Future of Humanity*.

2. Among contemporary writers, the work of Ronald Dworkin elaborates this idea with striking clarity and ingenuity. See generally, *Law's Empire* (1986) and *Taking Rights Seriously* (1977).

3. See Philip Alston, "A Third Generation of Solidarity Rights: Progressive Development or Obfuscation of International Human Rights Law," *Netherlands International Law Review*, vol. 29 (1982): 322.

4. For a biting critique of efforts to transform every sort of interest, but particularly those frequently included in lists of "third generation rights," into a human right, see Philip Alston, "Conjuring Up New Human Rights: A Proposal for Quality Control," *American Journal of International Law*, vol. 78 (1984): 607.

5. The call for a right to development can fairly be seen as another way of expressing the demand for a New International Economic Order. P.T. Bauer typifies the jaundiced reaction of Western conservatives: see, e.g., P.T. Bauer and B.S. Yamey, "Against the New Economic Order," *Commentary*, April 1977, p. 25.

6. For an on the whole positive response from Western writers, see I. William Zartman, ed., *Positive Sum* (Transaction Books, New Brunswick, 1987).

7. P.T. Bauer is without question the most effective proponent of the views summarized in this paragraph. See especially *Reality and Rhetoric* (Harvard Univerity Press, Cambridge, 1984) and *Dissent on Development* (Harvard University Press, Cambridge, 1976).

8. "Technology Growth in Human Rights Perspective," paper presented at the United Nations University's workshop on Human Rights and Scientific and Technological Development, Geneva, 16–18 November 1987 (hereinafter "workshop"), p. 15.

9. The Covenant came into effect on 23 March 1976; within ten years over 80 states had adhered. For a comprehensive guide to interpretation of the Covenant, see L. Henkin, ed., *The International Bill of Rights: The Convenant on Civil and Political Rights* (Columbia University Press, New York, 1981).

10. See, e.g., Richard Falk, "Nuclear Weapons and the Renewal of Democracy," in Cohen and Lee (note 1 above), *Future of Humanity*, p. 437.

11. For a very complete collection of the relevant documents, see Adam Roberts and Richard Guelff, eds., *Documents on the Laws of War* (Oxford University Press, Oxford, 1982).

12. The 1907 Hague Convention IX Concerning Bombardment by Naval Forces in Time of War would seem to be their earliest expression in an international agreement. But the distinction had been captured no later than the middle of the nineteenth century in national military manuals and rules of engagement.

13. See Michael Walzer, *Just and Unjust Wars* (Basic Books, New York, 1977), pp. 127–128.

14. Walzer (note 13 above), pp. 127–175.

15. International Military Tribunal (Nuremburg) Judgment and Sentences, *American Journal of International Law*, vol. 41 (1946): 172–175, 220–221.

16. "What are the implications of the just-war tradition for the use of nuclear weapons in war? The use of nuclear weapons in war violates both the principle of proportionality and the principle of discrimination" (Cohen and Lee (note 1 above)). And see generally pp. 231–372 in the same volume. See also Walzer (note 13 above), pp. 269–283. So profound is the challenge to traditional ethics posed by nuclear weapons that it drew the National Conference of Catholic Bishops into the unceasing national debate over their use; see their "Nuclear Strategy and the Challenge of Peace: Ethical Principles and Policy Prescriptions," in Charles W. Kegley Jr. and Eugene R. Wittkopf, eds., *The Nuclear Reader* (St Martin's Press, New York, 1985), p. 43. For an astringent critique of the bishops' position, see Albert Wohlstetter, "Bishops, Statesmen, and Other Strategists on the Bombing of Innocents," in Kegley and Wittkopf, p. 58.

17. For an elaboration of the discussion in the last three paragraphs, see generally Tom J. Farer, "The Laws of War Twenty-five Years after Nuremburg," *International Conciliation* (Carnegie Endowment, Washington, D.C., 1971).

18. In May 1955 five individuals instituted a legal action against the Japanese government to recover damages for injuries allegedly sustained as a consequence of the atomic bombing of Hiroshima and Nagasaki. On 7 December 1963, the District Court of Tokyo, while finding that the claimants had no legal basis for recovering damages from the Japanese government, concluded that the United States had violated international law by dropping atom bombs on Hiroshima and Nagasaki. See Richard Falk, "The Shimoda Case: A Legal Appraisal of the Atomic Attacks upon Hiroshima and Nagasaki," *American Journal of International Law*, vol. 59 (1965): 759.

19. "It was not until mid 1945, with the New Mexico test of the first bomb, that the full enormity of atomic power could be properly appreciated . . . In August 1941, after being informed that a weapon equivalent to 1,800 tons of TNT could be produced, Churchill noted his contentment with 'existing explosives' before recognizing that 'we must not stand in the path of improvement.'" Lawrence Freedman, *The Evolution of Nuclear Strategy* (St Martin's Press, New York, 1981), p. 16.

20. Freedman (note 19 above).

21. *The Just War: Force and Political Responsibility* (Scribners, New York, 1968).

22. See, in this connection, Robert S. McNamara's estimate of the impact on Western Europe alone of fighting involving the use of tactical nuclear weapons, in "The Military Role of Nuclear Weapons: Perceptions and Misperceptions," in Kegley and Wittkopf (note 16 above), p. 153.

23. See, e.g., Richard Pipes, "How to Cope with the Soviet Threat: A Long-term Strategy for the West," *Commentary*, August 1984, pp. 13–14.

24. See Henry Kissinger, *White House Years* (Little, Brown, Boston, 1979), p. 1250.

25. A cross-sectional sampling of the literature might include the following books and articles: Ashton B. Carter and David N. Schwartz, eds., *Ballistic Missile Defence* (Brookings Institution, Washington, D.C., 1984); *Daedalus*, vol. 114 (Summer 1985); Daniel O. Graham, *High Frontier: A New National Strategy* (Tor Books, New York, 1983); Fred Charies Ikie, "Nuclear Strategy: Can There Be a Happy Ending?" *Foreign*

Affairs, vol. 63 (1985): 824; Robert Jastrow, *How To Make Nuclear Weapons Obsolete* (Little, Brown, Boston, 1985); Union of Concerned Scientists, *The Fallacy of Star Wars* (Vintage Books, New York, 1984); Jonathan Stein, *H-Bomb to Star Wars* (Lexington Books, Lexington, Mass., 1984); James R. Schlesinger, "Rhetoric and Realities in the Star Wars Debate," *International Security*, vol. 10 (1985): 5.

26. The next generation Trident missile will, it is believed, have the accuracy now enjoyed exclusively by land-based missiles that makes them effective weapons against hardened targets like missile silos.

27. For a shrewd analysis of how arms control might affect the impact of SDI on the stability of the US–USSR strategic relationship, see Robert H. Gromoll, "SDI and the Dynamics of Strategic Uncertainty," *Political Science Quarterly*, vol. 102 (1987): 481, 496–499.

28. President Ronald Reagan, "Address to the Nation," *Weekly Compilation of Presidential Documents*, 28 March 1983, pp. 423–466; Jastrow (note 25 above).

29. See, e.g., the remarks of Richard DeLauer, former Under-Secretary of Defense for Research, Development, and Engineering in *Government Executive*, July–August 1983, and Donald Snow, *The Nuclear Future: Toward a Strategy of Uncertainty* (University of Alabama Press, 1983).

30. Gromoll (note 27 above).

31. Gromoll (note 27 above), pp. 489–491.

32. Gromoll (note 27 above), p. 495.

33. In preparing this section of my remarks, I have relied heavily on insights developed in a paper – "Privacy and Regulating the New Reproductive Technology: A Decision-making Approach" – written by my colleague at the University of New Mexico, Professor Antoinette Sedillo Lopez. Among other useful pieces in the available literature are the following: L. Andrews, *New Conceptions: A Consumer's Guide to the Newest Infertility Treatments Including In Vitro Fertilisation, Artificial Insemination and Surrogate Motherhood* (St Martin's Press, New York, 1984); Note, "Sex Selection Abortion: A Constitutional Analysis of the Abortion Liberty and a Person's Right to Know," *Indiana Law Journal*, vol. 56 (1981); John A. Robertson, "Procreative Liberty and the Control of Conception," *Virginia Law Review*, vol. 69 (1983): 405; Handel, "Surrogate Parenting, In Vitro Insemination and Embryo Transplantation," *Whittier Law Review*, vol. 6 (1984): 783; Kass, "Making Babies – The New Biology and the 'Old' Morality," *The Public Interest*, Winter 1972, p. 318; Note, "Redefining Mother. A Legal Matrix for New Reproductive Technologies," *Yale Law Journal*, vol. 96 (1986): 187.

34. Charles Krauthammer, "The Ethics of Human Manufacture," *The New Republic*, 4 May 1987, p. 17.

35. Tamar Lewin, *New York Times*, 16 August 1987, p. 16.

36. Krauthammer (note 34 above), p. 18.

37. Krauthammer (note 34 above), p. 17.

38. *Albuquerque Journal*, 13 December 1987, p. B-3.

39. *Albuquerque Journal* (note 38 above).

40. Richard Reeves, *Albuquerque Journal*, 5 October 1987, p. A–4.

41. Reeves (note 40 above).

42. *Civil Liberties*, Summer 1987, p. 3.

43. "Reviewing Privacy in an Information Society," *University of Pennsylvania Law Review*, vol. 135: 707.

44. "Reviewing Privacy. . ." (note 43 above), p. 709.
45. See note 43 above, p. 711.
46. See note 43 above, p. 712.
47. See note 43 above, p. 713.
48. See note 43 above, p. 713.
49. See note 43 above, p. 718.
50. See note 43 above, p. 718–719.

BIBLIOGRAPHY

Cohen, Avner, and Steven Lee. The Nuclear Predicament. In: A. Cohen and S. Lee, eds., *Nuclear Weapons and the Future of Humanity*. Rowman & Allanheld, Totowa, 1986.

Carter, Ashton B., and David N. Schwartz, eds. *Ballistic Missile Defense*. Brookings Institution, Washington, D.C., 1984.

Freedman, Lawrence. *The Evolution of Nuclear Strategy*. St Martin's Press, New York, 1981.

Jastrow, Robert. *How To Make Nuclear Weapons Obsolete*. Little, Brown, Boston, 1985.

Kegley, Charles W. Jr, and Eugene R. Wittkopf, eds. *The Nuclear Reader*. St Martin's Press, New York, 1985.

Union of Concerned Scientists. *The Fallacy of Star Wars*. Vintage Books, New York, 1984.

Waltzer, Michael. *Just and Unjust Wars*. Basic Books, New York, 1977.

Andrews, L. *New Conceptions: A Consumer Guide to the Newest Infertility Treatments Including In Vitro Fertilisation, Artificial Insemination and Surrogate Motherhood*. St Martin's Press, New York, 1984.

Part 3

International Response

5

Impacts of Scientific and Technological Progress on Human Rights: Normative Response of the International Community

HIROKO YAMANE

Since the adoption of the Universal Declaration of Human Rights in 1948, the international community has sought constantly to elaborate and refine the normative framework within which human rights should be respected. However, despite considerable discussion devoted to the subject, refinements of human rights legislation with respect to the world's rapidly advancing scientific and technological prowess have failed to keep pace with other branches of United Nations human rights law-making.

It is evident that, as societies evolve, so too do the conditions under which human rights in any given society can most effectively be realized. Those conditions naturally change and, in some cases, expand or become more complex. In addition, new scientific and technological developments also modify the means to act which the state, private enterprise, and the individual have at their disposal. As a result, both the nature and scope of human rights violations – as well as the conditions which must be met in order to protect those rights – are in a permanent state of flux and demand continual re-evaluation.

Advancements in science and technology have so far proved a mixed blessing with respect to the protection of human rights. For instance, cheaper and more efficient means of communication which have evolved over the past 20 to 30 years have served to increase the flow of information across borders of all kinds: geographical, political, industrial, and interpersonal. Thus the right to freedom of expression and, in particular, the right to information, is enhanced. However, it must be acknowledged that the same advances in technology that afford human beings greater access to information also permit governments, political parties, and other bureaucrats to gain even tighter control of that information. Furthermore, the relative ease with which information can now be communicated almost instantaneously to wide audiences heightens the dangers posed by the dissemination of faulty or distorted information, i.e. disinformation.

It is in these particular contexts that the slogan "the right to know" has become of crucial importance in interpreting Article 19 of the International Covenant on

Civil and Political Rights, which states: "Everyone shall have the right to freedom of expression; this right shall include freedom to seek, receive and impart information and ideas of all kinds . . ."

With the evolution of society, new conditions give birth to new aspirations which could lead to a reformulation of rights that have been expressed differently in previous situations.[1] However, the ease with which human rights are formulated and reformulated is in contrast to the absence of effective remedies. Contrary to common-law tradition, in which remedies precede rights, human rights are aspirations which can be declared even in the absence of remedies. The exercise in reformulating rights can foster the illusion that solutions can be found. Or worse, this could be a way to avoid providing solutions.

The international human rights norms related to science and technology offer a corpus of texts which can be analysed in terms of the underlying conceptions of man and society that prevail in the international community at a particular moment. Inasmuch as scientific and technological progress interferes with life and death, with different types of societies and communities (such as the family), as well as with nature and the environment, human rights problems which arise from this process are numerous and diverse. Varied discussions have developed internationally on such problems as the right to privacy, the beginning and end of life, manipulation of the mind, etc., from the human rights angle.

Despite the great amount of discussion on the impact of science and technology on human rights, the normative response resulting from it is relatively meagre. The international norms established to counter the effects of scientific and technological progress on human rights are often scattered among different kinds of other human rights problems such as torture and medical ethics, or find themselves as vague references in the declarations concerning welfare or development in general. The only international human rights instrument specifically related to scientific and technological progress is the Declaration on the Use of Scientific and Technological Progress in the Interest of Peace and for the Benefit of Mankind (10 November 1975), which expresses the wish that all states make use of scientific and technological progress for good purposes.

This study will retrace the evolution of the normative instruments which have been elaborated to cope with human rights problems faced with scientific and technological progress. Before doing so, however, a stocktaking of those human rights which are particularly affected by scientific and technological progress will be necessary.

HUMAN RIGHTS WHICH ARE PARTICULARLY AFFECTED BY SCIENTIFIC AND TECHNOLOGICAL PROGRESS

Among the human rights enumerated in the Universal Declaration of Human Rights, the following rights would seem to be particularly affected by scientific and technological progress:

- The right to life (Article 3), in the sense that science (biology, medicine, etc.) as well as technology (gene technology, nuclear technology,[2] etc.) can determine or influence birth and death. Problems posed by abortion, *in vitro* fertilization, embryo transplantation, euthanasia techniques, untested drugs, are examples.
- The right to physical and spiritual integrity (Article 5 stipulates that no one shall be subjected to torture or to cruel, inhuman, or degrading treatment or punishment). Use of drugs and other chemical controls of the mind, psychological and physical testing methods, and behaviour therapy are still often used in interrogation.
- The right to privacy (Article 12 stipulates that no one shall be subjected to arbitrary interference with his privacy, family, home, or correspondence, nor to attacks upon his honour and reputation . . .). The developments in recording, surveillance devices, personality tests, and other communication techniques based on electronics, optics, and acoustics, as well as new reproduction techniques, have considerably changed the ways in which privacy could be protected.
- The right to freedom of opinion and expression and the right to information (Article 19). Developments of micro-electronic communication technology have changed the conditions in which this right is exercised.

In a less precise way, the exercise of the following rights is influenced by scientific and technological progress:

- The right to property (Article 17).[3] Developments of new forms of property, such as software, have given rise to new thinking about the right to property.
- The right to work (Article 23). Developments of new technologies have changed market structures affecting the right to work.
- The right to a standard of living adequate for the health and well-being of himself and of his family, including food, clothing, housing, and medical care and necessary social services (Article 25). Scientific and technological progress can engender new forms of discrimination in the exercise of this right. Lack of access to medical information can also affect adversely the right to health.
- The right to education (Article 26). Developments in communication and information technology can promote this right but they can also create new forms of discrimination in education.
- The right freely to participate in the cultural life of the community, to enjoy the arts and to share in scientific advancement and its benefits (Article 27). Scientific and technological developments do not in themselves guarantee this right, but combined with the reinforcement of freedom of expression, the right to information and the right to education, this right can be promoted, thanks to a better communication technology.

Thus, scientific and technological progress can have both negative and positive impacts on human rights. The effects depend often on the right to information and freedom of expression. Inasmuch as technological progress produces conditions favourable to disinformation and cultural indoctrination, the right to information and freedom of expression seems to be a crucial factor in transforming scientific and technological progress into conditions conducive to a better respect of human rights.

The international norms so far developed to cope with human rights problems arising from scientific and technological progress seem to have ignored this crucial dimension – the right to information and freedom of expression – which inevitably intervenes when dealing with these problems.

INTERNATIONAL INSTRUMENTS AS A MEANS OF PROTECTION AGAINST THE ABUSE OF SCIENCE AND TECHNOLOGY

Instruments of a General Character

The provisions contained in the International Covenants on Economic, Social, and Cultural Rights, as well as on Civil and Political Rights, adopted on 16 December 1966, reiterate those rights enumerated above.

The Covenant on Civil and Political Rights protects the right to life (Article 6), the right to physical and spiritual integrity (Article 7), the right to privacy (Article 17), and the right to information (Article 19). Article 7 stipulates specifically that "no one shall be subjected without his free consent to medical or scientific experimentation." Article 19 adds details about various forms of communication for receiving and imparting information, implying that freedom of expression should be adapted to the conditions posed by the advances in communication technology. ("Everyone shall have the right to freedom of expression; this right shall include freedom to seek, receive and impart information and ideas of all kinds, regardless of frontiers, either orally, in writing or in print, in the form of art, or through any other media of his choice.")

Instruments of a Specific Character

The Right to Life

The Convention on the Prevention and Punishment of the Crime of Genocide, adopted by the General Assembly on 9 December 1948, protects in Article 11(d) the right to life against the abuse of science and technology:

In the present Convention, genocide means any of the following acts committed with intent to destroy, in whole or in part, a national, ethnical, racial or religious group as such:
(a) killing members of the group;
(b) causing serious bodily or mental harm to members of the group;
(c) deliberately inflicting on the group conditions of life calculated to bring about its physical destruction in whole or in part;
(d) imposing measures intended to prevent births within the group;
(e) forcibly transferring children of the group to another group.

The problems relating to the right to life posed by recent developments in gene technology have inspired many international and national research projects on this matter.[4] The Council of Europe, as well as Unesco, have ongoing research programmes on the impacts of gene technology on human rights. However, this has not yet reached the stage of a normative instrument being drafted.

The Right to Physical and Spiritual Integrity

Although they do not specifically refer to medical, scientific, or biological techniques, the Declaration on the Protection of All Persons from Being Subjected to Torture and Other Cruel, Inhuman or Degrading Treatment or Punishment (9 December 1975), the Code of Conduct for Law Enforcement Officials (17 December 1979), as well as the Convention against Torture and Other Cruel, Inhuman or Degrading Treatment or Punishment (10 December 1984) protect this right against torture.

The Principles of Medical Ethics relevant to the Role of Health Personnel, particularly Physicians, in the Protection of Prisoners and Detainees against Torture and Other Cruel, Inhuman or Degrading Treatment or Punishment (18 December 1982) explicitly stipulates in Principle 4(a):

It is a contravention of medical ethics for health personnel, particularly physicians,
(a) to apply their knowledge and skills in order to assist in the interrogation of prisoners
   and detainees in a manner that may adversely affect the physical or mental health or
   condition of such prisoners or detainees and which is not in accordance with the relevant international instruments; . . .

Within the Council of Europe, the Convention for the Prevention of Torture and Inhuman or Degrading Treatment or Punishment was adopted on 7 July 1987, establishing a European Committee for the Prevention of Torture and Inhuman or Degrading Treatment. The Committee is authorized by the state parties to visit those persons who are deprived of their liberty by public authorities.

The Right to Privacy

National legislation has been elaborated in many Western countries to protect personal data (Sweden in 1973, United States in 1974, New Zealand in 1976, Federal Republic of Germany in 1977, Denmark in 1978, Norway in 1978, France in 1978, Canada in 1982 and Japan in 1988). Within the United Nations, however, no specific normative instrument has been drawn up to protect the right to privacy. In the OECD and the Council of Europe, the development of computer-telecommunications technology gave rise to a movement to protect the right to privacy, especially in respect of the handling of personal data. The OECD adopted in 1981 the "Guidelines on the Protection and Transborder Flows of Personal Data." The Council of Europe adopted, also in 1981, the "Convention for the Protection of Individuals with regard to Automatic Processing of Personal Data." It protects the right to privacy in the automatic processing of personal data, without prohibiting their transborder flows.

Unesco has also encouraged research in comparative legislation concerning data protection, although no international norms have been established. The initial research result covering 1968–1971 was published in the *Social Science Journal* in 1972.[5] Unesco updated this study and added some intercultural research on the notion of privacy for publication in 1988.

The Right to Information

In 1947 the United Nations established a Sub-Commission on Freedom of Information and the Press. A world conference on this subject was held in 1948, but the Sub-Commission ceased to function in 1952, due in part to controversies over its functions. As a consequence, no normative instrument has been elaborated by the UN on the right to information, with the exception of the Convention on the International Right of Correction (16 December 1952). This Convention assures the contracting state the right to exercise the right of correction against the contracting states within whose territories a news despatch capable of damaging the state's prestige or dignity or its relations with other states has been published or disseminated. Despite the research effort made at the UN on the impacts of new information and communication technologies on human rights, no attempt has been made to draw up international instruments to reinforce the right to obtain the information held by state bureaucracies, local administration, or institutions with a public mission (e.g. schools, scientific research institutes, nuclear power plants, etc.). Recently, significant developments have been observed in the national laws ensuring the access to information or documents held by the administration or the institutions carrying out public missions (Freedom of Information Act in the United States in 1966 amended in 1976, Denmark in 1970, Norway in 1970, the Netherlands in 1978, France in 1978, Canada in 1982, Australia in 1982, New Zealand in 1982, United Kingdom in 1985). In comparison with these developments in national laws, no attempt has been made to elaborate an international instrument to protect this "right to know" vis-à-vis public or semi-public institutions. From the point of view of the effects of scientific and technological progress on human rights, this seems to be the most important right, in terms of both the positive and the negative effects of such progress. At the national level, however, information related to the effects of industrial waste on water, the side-effects of medical products on human bodies, or the content of pollution due to power plants or uranium recycling plants can be requested by virtue of legislation allowing access to information or administrative documents.

Within the Council of Europe, attempts have been made to draw up a Convention on mass media based on Article 10 of the European Convention on Human Rights (4 November 1950), which stipulates: "Everyone has the right to freedom of expression. This right shall include freedom to hold opinions and to receive and impart information and ideas without interference by public authority and regardless of frontiers . . ."

At the Colloquium organized in Sevillia in November 1986, which examined this right in connection with the restrictions "necessary in a democratic society"[6] contained in paragraph 2 of the same Article, it was argued that such a new formula as the right to know or the right to have access to information, which is not explicitly recognized in the text of the Convention, could be included in the right to freedom of expression, if it affirmed in the case law of the European Convention. The right of the public to receive information and the duty of the mass media to contribute to it, affirmed by the European Court of Human Rights in the *Sunday Times* case (judgment of 26 April 1979) is a case in point. This case is of particular interest, reflecting as it does the impacts of scientific progress on human rights, as it concerns the harmful side-effects of medicine (thalidomide). In this case, Article 10 of the European Convention on Human Rights was interpreted to include the right of the public to "know" the effects of pharmaceutical products and the obligation of the mass media to diffuse such information.

INSTRUMENTS AS A MEANS TO ASSURE POSITIVE USES OF SCIENTIFIC AND TECHNOLOGICAL PROGRESS FOR ADVANCING HUMAN RIGHTS

Instruments of a General Character

The Covenant on Economic, Social, and Cultural Rights refers vaguely to the obligation of the state to use scientific and technological progress for welfare. For example, to achieve the full realization of *the right to work*, Article 6, para. 2 stipulates that the state should take steps including "technical and vocational guidance and training programmes, policies and techniques to achieve steady economic, social and cultural development and full and productive employment under conditions safeguarding fundamental political and economic freedoms to the individual."

The state should also take the following steps:

– To improve methods of production, conservation and distribution of food by *making full use of technical and scientific knowledge, by disseminating knowledge of the principles of nutrition* and by developing or reforming agrarian systems in such a way as to achieve the most efficient development and utilization of natural resources. (Article 11, para. 2(a))
– [To ensure the] improvement of all aspects of environmental and industrial hygiene in order to realize *the right of everyone to the enjoyment of the highest attainable standard of physical and mental health*. (Article 12, para. 2(b))

Much state action is expected to provide welfare, but few mechanisms have been developed to oblige the appropriate authorities to take such action. On the other hand, the question of private enterprise is not dealt with, perhaps because it touches delicately upon the important premises of liberal-individualistic society: the market economy, freedom of private enterprise, the right to property, freedom of expression, publicity, etc.

The Covenant on Economic, Social, and Cultural Rights also assures the individual the right to enjoy the benefits of scientific progress and its applications (Article 15, para. 1(b)).

Instruments of a Specific Character

The Right to Benefit from Scientific Progress

The Unesco Recommendation on the Status of Scientific Researchers, adopted on 20 November 1974, combines the problem of the freedom of researchers and the implications of science and technology for world problems such as development and international peace.

This Recommendation recalls the Universal Declaration of Human Rights, particularly its Article 27, which provides that everyone has the right freely to participate in the cultural life of the community and to share in scientific advancement and its benefits. It recognizes, *inter alia*, that "A cadre of talented and trained personnel is the cornerstone of an indigenous research and experimental development capability and indispensable for the utilization and exploitation of research carried out elsewhere."

It provides under chapter IV, the Vocation of the Scientific Researcher, that:

Member states should seek to encourage conditions in which scientific researchers, with the support of the public authorities, have the responsibility and the right
(a) to work in a spirit of intellectual freedom to pursue, expound and defend the scientific truth as they see it;
(b) to contribute to the definition of the aims and objectives of the programmes in which they are engaged and to the determination of the methods to be adopted which should be humanely, socially, and ecologically responsible;
(c) to express themselves freely on the human, social, or ecological value of certain projects and in the last resort withdraw from those projects if their conscience so dictates;
(d) to contribute positively and constructively to the fabric of science, culture, and education in their own country, as well as to the achievement of national goals, the enhancement of their fellow citizens' well-being, and the furtherance of the international ideals and objectives of the United Nations."

The Right to an Adequate Standard of Living

The Proclamation of Tehran, adopted by the International Conference on Human Rights on 13 May 1968 to celebrate the twentieth anniversary of the Universal Declaration of Human Rights, introduced the distinct categories of developed and developing countries with regard to human rights. Article 12 of the Proclamation affirms:

The widening gap between the economically developed and developing countries impedes the realization of human rights in the international community. The failure of the Development Decade to reach its modest objectives makes it all the more imperative for every nation, according to its capacities, to make the maximum possible effort to close this gap.

The same proclamation calls attention to the positive and negative effects of scientific and technological progress on human rights, with the emphasis on the potential for development offered by such progress:

While recent scientific discoveries and technological advances have opened vast prospects for economic, social and cultural progress, such developments may nevertheless endanger the rights and freedoms of individuals and will require continuing attention. (Article 18)

This distinction between developing and developed states having been introduced into the categories of states, much attention has gradually turned to the rights and responsibilities of the state in exercising its economic and social functions for spreading welfare to developing countries. This has initiated a period where human rights are intricately intertwined with the rights and duties of the state and the rights of peoples, as well as all-inclusive issues of development. The role of information science and technology was also to be associated with human rights, viewed from the particular angle of development.

The Declaration on the Use of Scientific and Technological Progress in the Interests of Peace and for the Benefit of Mankind, adopted by the General Assembly of the United Nations on 10 November 1975, takes into consideration that, "while scientific and technological developments provide ever-increasing opportunities to better the conditions of life of peoples and nations, in a number of instances they can give rise to social problems, as well as threaten the human rights and fundamental freedom of the individual." This Declaration proclaims in all its nine articles the responsibilities of the state, particularly

- to promote international co-operation to use the results of scientific and technological developments for strengthening international peace and security and freedom and independence, and for the purpose of the economic and social development of peoples and the realization of human rights;
- to prevent scientific and technological development from being used to curtail the enjoyment of human rights;
- to co-operate in increasing the scientific and technological capacity of developing countries with a view to accelerating the realization of the social and economic rights of the peoples of those countries;
- to extend the benefits of science and technology to all strata of the population and to protect them from possible harmful effects of the misuse of scientific and technological developments.

This Declaration indicates that the major preoccupation of the United Nations in dealing with the impact of science and technology on human rights has turned to economic development and that, in order to achieve this goal, more and more emphasis has been placed on the economic and social functions of the state, both internally and externally.

This tendency towards the identification of a connection between human rights and development coincided with the ongoing effort of the international community to establish a new international economic order. The Declaration on the Establishment of a New International Economic Order, adopted by the General Assembly on 1 May 1974, proclaims the "united determination to work urgently for the establishment of a new international economic order based on equity, sovereign equality, interdependence, common interest, and co-operation among all states, irrespective of their economic and social systems, which shall correct inequalities and redress existing injustices, make it possible to eliminate the widening gap between the developed and the developing countries. . . ." Article 4(p) of the Declaration states the principle of "Giving to the developing countries access to the achievements of modern science and technology, and promoting the transfer of technology and the creation of indigenous technology for the benefit of the developing countries in forms and in accordance with procedures which are suited to their economies." This is followed by two other principles:

(q) the need for all states to put an end to the waste of natural resources, including food products;
(r) the need for developing countries to concentrate all their resources for the cause of development.

The overriding concern with development gives new rights to the *state*. The Charter of Economic Rights and Duties of States adopted by the General Assembly in the same year grants the *state* the right, *inter alia*
− to regulate and supervise the activities of transnational corporations within its national jurisdiction (Article 2, para. 2(b)); and
− to benefit from the advances and developments in science and technology for the acceleration of its economic and social development (Article 13, para. 1).
The Charter stresses, at the same time, the duties of the "developed countries to co-operate with the developing countries in the establishment, strengthening, and development of their scientific and technological infrastructures and their scientific research and technological activities so as to help to expand and transform the economies of developing countries" (Article 13, para. 3), and the duties of "all states to co-operate in research with a view to evolving further internationally accepted guidelines or regulations for the transfer of technology, taking fully into account the interests of developing countries" (Article 13, para 4).

The UN Conference on Science and Technology for Development (UNCSTD), held in Vienna in 1979, adopted the same line of thought, in a Declaration encouraging future programmes to be conducted for exploring alternative technologies and a better use of science and technology for development.

The implementation of human rights requires appropriate social conditions in all aspects of society. It is therefore natural that the conditions allowing for the implementation of the individual right to benefit from scientific and technological progress, as well as the right to an adequate standard of living, include not only the change of national social conditions, but also the entire world order.

It was in a similar perspective that Unesco carried out, pursuant to General Assembly resolution 33/115 of 18 December 1978, consultations on ways and means by which assistance for developing countries could be increased in the field of communication technology and systems for their social progress and economic development. Subsequently, the Declaration on Fundamental Principles concerning the Contribution of the Mass Media to Strengthening Peace and International Understanding, to the Promotion of Human Rights, and to Countering Racism, Apartheid, and Incitement to War was adopted on 28 November 1978.

This Declaration, which aims at creating a "new international information order," conceives, as in other instruments related to the establishment of the new international economic order, of human rights primarily in the context of the elimination of disparities that exist between the developed and developing countries and of the liberation of oppressed peoples. Mass media in this Declaration are considered as a means for achieving human rights, conceived of as a social order to be realized. Mass media, whether in the form of private or state enterprise, have responsibilities to discharge in establishing this order. The meaning of human rights thus loses its individualistic character and becomes action to establish an international order. For those who consider freedom of expression of the private media to be an essential part of human rights, this Declaration, expressing national aspirations for economic development, does not really deal with the right to information as such. Economic well-being is certainly crucial to the exercise of human rights. A clash of principles and interests seems inevitably to occur on the question of whether the liberal–individualistic approach to human rights is compatible with the approach which calls for the active intervention of the state in many spheres of life, from childhood to old age, in education, culture and the economy, as well as in establishing a new international economic order – all these state actions together constituting a *conditio sine qua non* for the respect of human rights.

Once the international debate on human rights opens up all these divisive questions concerning the relationship between the state and the individual, more effort seems to be spent on arriving at superficial conceptual compromises than on devising effective ways and means of implementing human rights. One result of merely claiming and proclaiming rights of all kinds together can be the lack of conceptual clarities that we encounter in the international normative texts, as well as the impossibility of achieving real progress in devising remedies.

One of the possible solutions to this impasse would be to specify the areas for implementing different categories of human rights and to dissociate them from the too general issues of economic development, such as transfer of technology, exploitation of natural resources, etc. These latter areas concern primarily the domain of the social and economic functions of the state, private enterprise, the economy and the settlement of disputes between private enterprise and the state.

These problems could perhaps be better solved if they were dealt with in their proper context.[7]

However, the search for remedies through specification of contexts and issues does not seem to be fashionable today. The Declaration on the Right to Development, adopted by General Assembly resolution 41/128 on 4 December 1986, is a case in point. This Declaration reiterates all the civil and political, as well as economic, social, and cultural rights contained in the Covenants to encourage international co-operation to enable "more rapid development of developing countries" (Article 4, para. 2). The adoption of this instrument could be of help in attenuating North–South verbal conflicts,[8] but it hardly provides any real remedies.

POSSIBLE AREAS FOR DRAWING UP NEW INSTRUMENTS RELATED TO THE EFFECTS OF SCIENTIFIC AND TECHNOLOGICAL PROGRESS ON HUMAN RIGHTS

The above overview of international instruments relating to scientific and technological progress demonstrates clearly that the process of norm-making in this field is developing only haphazardly and is not providing sufficient and effective remedies for the problems at hand. Many declarations have been made which urge that scientific and technological advances be put to the best possible use for the greater social good, and few people would argue with these sentiments. But while such declarations do play an important role in sensitizing public consciousness to the human rights issues at hand, they are neither designed to be nor capable of acting as surrogate normative guidelines by which these human rights can be exercised.

The crucial issue in the process of finding an equilibrium between the progress of science and technology and the protection of human rights centres on a single question: How does the control of information conflict with the fundamental human right to that information? If, for reasons of state secrecy, national defence, public security, or the protection of industrial secrets, the public is denied access to information concerning the *adverse* effects of science and technology, it is conceivable that the rights to free expression, health, safety, and participation in democratic life could be threatened. National laws designed to ensure free access to information held by public institutions all contain exceptions. Among these exceptions – most of which are legitimate – are information related to defence, sensitive diplomatic negotiations, and public security, as well as information protected by industrial property laws and personal privacy laws. But these exceptions were meant to be just that – exceptions – and must not be abused. In particular, when human rights and any potential violation of these rights are at stake, institutional mechanisms by which information can be requested should be installed and maintained within any given system of information processing.

There is no question that human beings have the right to benefit from scientific and technological progress. And this right logically implies that in order to make responsible qualitative judgments on how this "benefit" should be defined, people have the right to information concerning scientific and technological developments. Again it is clear that the key principle involved in making decisions about human rights in a world of rapidly advancing science and technology is "the right to information." As scientific and technological developments continue to play an increasingly significant role in more and more human lives, the right to have free access to accurate, truthful, and complete information concerning these developments should be self-evident. It is thus appropriate to elaborate an international instrument which will reinforce the right to information. The absence of information equals, eventually, the absence of real choice. And for democratic societies, nothing is more dangerous.

NOTES

1. See, for example, *Review of the Multi-lateral Treaty-making Process*, UN Soc. ST/LEG/ SER.B/21 (1982); Philip Alston, "Conjuring Up New Human Rights: A Proposal for Quality Control," *American Journal of International Law*, vol. 78 (1984): 607; Theodor Mern, "Reform of Lawmaking in the United Nations: The Human Rights Instance," *American Journal of International Law*, vol. 79 (1985): 664.
2. The Human Rights Committee included in the general comments on Article 6 of the Covenant on Civil and Political Rights that the "designing, testing, manufacture, possession, and deployment of nuclear weapons are among the greatest threats to the right to life," and that "use of nuclear weapons should be prohibited and recognized as crimes against humanity." *Report of the Human Rights Committee: General Assembly Official Records: Fortieth Session Supplement*, no. 40 (A/40/40), pp. 162–163.
3. In comparison to the Universal Declaration on Human Rights, the Convenants do not include the right to property. But the Covenant on Economic, Social, and Cultural Rights protects the right to intellectual and industrial property (Article 15, 1(c)).
4. See, for example, *Génétique, procréation et droit*, Proceedings of the Colloquium organized by the Comité Consultatif National d'Ethique pour les Sciences de la Vie et de la Société (Actes Sud, Arles, 1985), which deals mainly with the effects of genetic engineering on human rights.
5. *International Social Science Journal*, vol. 24, no. 3 (1972).
6. Final Report for the 6th International Colloquium on the European Convention on Human Rights, Sevillia, 13–16 November 1986 (Council of Europe, H/Coll(85)17).
7. Thus, the draft code of conduct on the transfer of technology (TC/CODE TOT/47) does not refer to human rights, but only to the principle that "technology is key to the progress of mankind and all peoples have the right to benefit from the advances and developments in science and technology in order to improve their standards of living" (Preamble, para. 2). It is addressed to parties including "any person, either natural or juridical, of public or private law, either individual or collective, such as corporations, companies, firms, partnerships and other associations, or any combination thereof, whether created, owned or controlled by states, government agencies, juridical persons, or individuals, wherever they operate, as well as states, government agencies and

international, regional, and subregional organizations, when they engage in an international transfer of technology transaction which is usually considered to be of a commercial nature" (chap. I: Definitions and Scope of Application, para. 1.1(a)).

It is within this delimited, specialized field that the draft code delineates objectives and principles, among which: "facilitating and increasing the access to technology, particularly for developing countries, under mutually agreed fair and reasonable terms and conditions, are fundamental elements in the process of technology transfer and development" (chap. 11.2: Principles (vii)).

Evidently, the conflict of interests and the different positions of the states make the adoption of this text difficult. Still, the issues are not of a rhetorical nature, which makes it possible to explore the ways and means of strengthening the bargaining power of those who are in a weak negotiating position.

8. Summary Record of the 61st meeting (28 November 1986) of the Third Committee, United Nations, A/C.3/41/SR.61 pf 5 December 1986.

BIBLIOGRAPHY

United Nations. *Human Rights: A Compilation of International Instruments*. 1988. (ST/HR/1/ Rev 3.)

6

The Institutional Response[1]

YO KUBOTA

INTRODUCTION

Universal awareness of the fact that scientific and technological developments do not occur in a vacuum but that they are inextricably linked to and affect human rights has never been more acute than in the latter half of the twentieth century. This consciousness has understandably been reflected in the work of the United Nations Human Rights organs.

The Universal Declaration of Human Rights of 1948 was written in the atmosphere of the aftermath of the Second World War, when the full extent of the atrocities committed by the hand of or in the name of scientific development were made known and there was a strong impetus for the promotion and protection of human rights. More recently, the far-reaching advances in information technology, biotechnology, and arms technology, among others, and their impact on human rights have also been the subject of debate and study in the United Nations human rights fora.

It is precisely man's thirst for knowledge and understanding that has led not only to scientific and technological developments but also to the need to study the consequences of progress in these fields.

At its forty-second session in 1986, tbe United Nations Commission on Human Rights adopted resolution 1986/9, entitled "Use of Scientific and Technological Developments for the Promotion and Protection of Human Rights and Fundamental Freedoms"[2] under agenda item 15, "Human Rights and Scientific and Technological Developments." The original draft of this resolution, contained in document E/CN.4/1986/L.59, was sponsored by Japan and Yugoslavia and introduced by Japan.

Introducing the draft resolution, the representative of Japan, Ambassador Tomohiko Kobayshi, stated that "the goal of the draft was to invite the United Nations University to study both the positive and the negative impact of scientific and technological developments on human rights."[3] Ambassador Kobayshi had earlier taken the floor and expressed his views on the question in more detail, stating his country's intention to support a research project on the relationship between human rights and scientific and technological progress by the United Nations University:

Mr Kobayshi (Japan) said that, over the years, the United Nations had undertaken a number of useful studies which had helped to alert the international community to the negative impact of scientific and technological developments on human rights. It might, indeed, be desirable to make an overall review of the studies that had been carried out, in the light of recent advances, with a view to updating them and taking any further action that might be necessary.

Regarding the positive effects of a proper use of science and technology, the comments submitted by states in response to Commission resolutions 1983/41 and 1984/27 (E/CN.4/ 1986/27 and 28) showed how effective science and technology could be in promoting human rights and fundamental freedoms. Of course, as many states had noted, the positive and negative effects of scientific and technological developments on human rights were intimately interrelated, and constituted, as it were, a double-edged sword. In considering the relationship between scientific and technological developments and human rights, it should not be forgotten that the former did not necessarily constitute a prerequisite for or a guarantee of the latter. There were many technologically advanced countries in which the basic human rights and fundamental freedoms of the individual were not fully realized.

A number of academic institutions were engaged in research on the relationship between scientific and technological progress and human rights. The United Nations University, for instance, was planning a preliminary study on the negative and positive impact of scientific and technological developments on human rights and fundamental freedoms, and his government intended to make a financial contribution to that project in the hope that other governments would do likewise.[4]

At the twenty-second meeting of the same session, the co-sponsor of the draft resolution, the representative of Yugoslavia, stated the following:

Mrs Djordjevic (Yugoslavia) recalled that science and technology had become a subject for consideration by the international community in the early 1970s and that the interrelationship between human rights and scientific and technological development had first received international recognition at the 1968 Tehran Conference. Since that time, the Commission had conducted many studies and adopted many resolutions on the issue.

However, the negative effects on human rights had always been emphasized, and an effort should, consequently, be made to review the many benefits which scientific and technological developments could have for the promotion of human rights and fundamental freedoms. That concept underlay Commission resolutions 1983/41 and 1984/27, submitted jointly by the Yugoslav and Japanese delegations some years earlier. A broad approach should be adopted, as a part of wide-ranging efforts to place material development at the service of man.

Her delegation was grateful to all those Member States and international organizations which, in compliance with the above-mentioned resolutions, had submitted to the Secretary-General their views on the most effective ways and means of using the results of scientific and technical developments for the promotion of human rights. However, the relationship between science and technology and human rights was one that evolved constantly, and the Commission should follow up developments, with the assistance of the Sub-Commission and other organizations, including academic institutions. Her delegation therefore supported the research planned by the United Nations University on the impact, both positive and negative, of science and technology on human rights, and hoped that the study would further stimulate the deliberations on the issue at the Commission's forty-fourth session.[5]

In operative paragraph 3 of the adopted resolution, numbered 1986/9, the Commission on Human Rights invited the United Nations University, in co-operation with other interested academic and research institutions, to study both the positive and negative impacts of scientific and technological developments and expressed the hope that the United Nations University would inform the Commission on Human Rights of the results of its own study on the question.[6]

In response to the above invitation, the United Nations University has decided to undertake such a study.[7] The general objective of the project is to develop a conceptual framework which will enable the discernment of both negative and positive impacts of scientific and technological developments on human rights and fundamental freedoms.[8]

The present article has been written in response to the United Nations University's invitation. The author of this article intends to consider the discussion to date within the United Nations human rights bodies, particularly in the Commission on Human Rights and its Sub-Commission on Prevention of Discrimination and Protection of Minorities, of the critical issues surrounding the relationship between human rights and scientific and technological developments, and to offer possible directions for its future study.

INITIAL STEPS

The promotion and protection of human rights and fundamental freedoms have been affected both positively and adversely by scientific and technological developments. Explaining this point, one Member State (the Federal Republic of Germany) stated in its submission to the Commission, that: "Scientific and technological progress have laid highly important foundations for the realization of human rights. Nevertheless, the fact cannot be ignored that technological developments have and will no doubt continue to have consequences which affect respect for human rights."[9]

Equally, an author from the German Democratic Republic expressed his views on this matter in the following way:

Technological advancements have always been potential means in man's hand, and they have always served human requirements. And it is man who has to decide the ends to which science and technology shall be used. And as man is a social being, his steps taken in this regard reflect necessarily the advantages or inadequacies in the political, social, and intellectual organization of society.[10]

On reflection, therefore, the Universal Declaration of Human Rights, the first international human rights instrument, and the International Covenants on Human Rights, the first international human rights conventions of a general and universal character, as might be expected, all contain provisions relating to various aspects of the effect of scientific and technological developments. However, this question was not considered in detail until 1968, when it was the subject of debate at the International Conference on Human Rights in 1968.[11]

In paragraph 18 of the Proclamation of Tehran, adopted by the Conference on 13 May 1968, the Conference expressed the view that "While recent scientific discoveries and technological advances have opened vast prospects for economic, social, and cultural progress, such developments may nevertheless endanger the rights and freedoms of individuals and will require continuing attention." It later adopted the resolution on "human rights and scientific and technological developments."

On the basis of the above-mentioned resolution, the General Assembly, in its resolution 2450 (XXIII) of 19 December 1968, invited the Secretary-General to undertake, with the assistance, *inter alia*, of the Advisory Committee on the Application of Science and Technology to Development, and in co-operation with the executive heads of the competent specialized agencies, a study of the problems in connection with human rights arising from developments in science and technology, in particular from the following respects:

(a) Respect for the privacy of individuals and the integrity and sovereignty of nations in the light of advances in recording or other techniques.
(b) Protection of the human personality and its physical and intellectual integrity in the light of advances in biology, medicine, and biochemistry.
(c) Uses of electronics which may affect the rights of the person and the limits which should be placed on such uses in a democratic society.
(d) More generally, the balance which should be established between scientific and technological progress and the intellectual spiritual, cultural, and moral advancement of humanity.

The General Assembly, in resolution 2480 (XXIII), also requested the Secretary-General "to prepare, on a preliminary basis, a report comprising a summary account of studies already made or in progress on the aforementioned subjects, emanating in particular from governmental and intergovernmental sources, the specialized agencies and the competent non-governmental organizations, and a draft programme of work which might be undertaken in fields in which subsequent surveys would be necessary for the attainment of the objectives of the resolution."

The preliminary version of the report, requested by the General Assembly, was available to the Commission on Human Rights at its twenty-sixth session in

1970[12] for consideration. However, the Commission did not have sufficient time to study the substance of the report in depth at the session.

The Commission consequently made a thorough examination of the question at its twenty-seventh session in 1971, and in its resolution 10 (XXVII), recognizing the need to study further, decided to make a series of requests addressed to different entities to initiate or continue studies on the relationship between science and technological developments and human rights.[13] It was not until this resolution 10 (XXVII) of 18 March 1971 that the Commission recognized the need during the Second United Nations Development Decade to concentrate its attention on the most important and basic problems of protecting human rights and fundamental freedoms in the context of scientific and technological progress, and in particular on:

(a) Protection of human rights in the economic, social, and cultural fields in accordance with the structure and resources of states and the scientific and technological level that they have reached, as well as protection of the right to work in conditions of the automation and mechanization of production.

(b) The use of scientific and technological developments to foster respect for human rights and the legitimate interests of other peoples and respect for generally recognized moral standards and standards of international law.

(c) Prevention of the use of scientific and technological achievements to restrict fundamental democratic rights and freedom.

AN OUTLINE OF DEVELOPMENTS, STUDIES, AND REPORTS, 1971–1987

Developments: The Declaration of the Use of Scientific and Technological Progress in the Interests of Peace and for the Benefit of Mankind

Attention must be directed to the Declaration of the Use of Scientific and Technological Progress in the Interests of Peace and for the Benefit of Mankind, as it serves to demonstrate the direction of UN activities in the relationship between human rights and scientific and technological developments, in the 1970s and the first half of the 1980s.

At its twenty-ninth session, in 1974, the General Assembly considered briefly a draft declaration on the use of scientific and technological progress in the interests of peace and for the benefit of mankind, submitted by Bangladesh, the Byelorussian Soviet Socialist Republic, the German Democratic Republic, Hungary, Mauritius, Poland, and the Union of Soviet Socialist Republics.[14] At its thirtieth session, in 1975, the Assembly considered a revised declaration.[15]

In resolution 3384 (XXX) of 10 November 1975, the General Assembly proclaimed the Declaration on the Use of Scientific and Technological Progress in the Interests of Peace and for the Benefit of Mankind. It took into account that, while scientific and technological developments provided ever-increasing opportunities to better the conditions of life of peoples and nations, in a number of instances they could give rise to social problems as well as threaten the human rights and fundamental freedoms of the individual, and proclaimed the following:

1. All states shall promote international co-operation to ensure that the results of scientific and technological developments are used in the interests of strengthening international peace and security, freedom and independence, and also for the purpose of the economic and social development of peoples and the realization of human rights and freedoms in accordance with the Charter of the United Nations.

2. All states shall take appropriate measures to prevent the use of scientific and technological developments, particularly by the state organs, to limit or interfere with the enjoyment of the human rights and fundamental freedoms of the individual as enshrined in the Universal Declaration of Human Rights, the International Covenants on Human Rights and other relevant international instruments.

3. All states shall take measures to ensure that scientific and technological achievements satisfy the material and spiritual needs of all sectors of the population.

4. All states shall refrain from any acts involving the use of scientific and technological achievements for the purpose of violating the sovereignty and territorial integrity of other states, interfering in their international affairs, waging aggressive wars, suppressing national liberation movements or pursuing a policy of racial discrimination. Such acts are a flagrant violation of the Charter of the United Nations and of the principles that should guide scientific and technological developments for the benefit of mankind.

5. All states shall co-operate in the establishment, strengthening, and development of the scientific and technological capacity of developing countries with a view to accelerating the realization of the social and economic rights of the peoples of those countries.

6. All states shall take measures to extend the benefits of science and technology to all strata of the population and to protect them, both socially and materially, from possible harmful effects of the misuse of scientific and technological developments, including their misuse to infringe upon the rights of the individual or of the group, particularly with regard to respect for privacy and the protection of the human personality and its physical and intellectual integrity.

7. All states shall take the necessary measures, including legislative measures, to ensure that the utilization of scientific and technological achievements promotes the fullest realization of human rights and fundamental freedoms without any discrimination whatsoever on grounds of race, sex, language, or religious beliefs.

8. All states shall take effective measures, including legislative measures, to prevent and preclude the utilization of scientific and technological achievements to the detriment of human rights and fundamental freedoms and the dignity of the human person.

9. All states shall, whenever necessary, take action to ensure compliance with legislation guaranteeing human rights and freedoms in the context of scientific and technological developments.

Following the adoption of the Declaration in 1975, the General Assembly, in resolution 31/128 of 16 December 1976, expressed its concern at the fact that scientific and technological achievements might be used to the detriment of fundamental human rights and fundamental freedoms, the dignity of the human person, international peace and security and social progress, and called upon Member States to take accournt of the provisions and principles contained in the Declaration on the Use of Scientific and Technological Progress in the Interests of Peace and for the Benefit of Mankind. The Assembly requested the specialized agencies concerned to take the provisions of the Declaration fully into account in their programme and activities, and requested the Commission on Human Rights, in its consideration of the question of scientific and technological progress and human rights, to give special attention to the implementation of the provisions of the Declaration.

The implementation activities of the Declaration on the Use of Scientific and Technological Progress in the Interests of Peace and for the Benefit of Mankind were also undertaken by the Commission on Human Rights at the request of the Assembly.

Five years after the adoption of the Declaration of 10 November 1975, in 1980, the General Assembly adopted resolution 35/130A, concerning the implementation of the provisions of the Declaration. In accordance with requests in the resolution made by the Assembly, the Secretary-General prepared reports on a variety of relevant issues, including problems for human rights caused by advances in biology, medicine and biochemistry, and those relating to human experimentation and to genetic manipulation. Studies relating to national and international machinery on technological assessments, needed to ensure that the short-term and long-term effects of new developments are not detrimental to human rights, had also been undertaken. Studies on these aspects were also prepared by the Commission on Human Rights and the Sub-Commission on Prevention of Discrimination and Protection of Minorities.

At its thirty-seventh session in 1982, the General Assembly, in its resolution 37/189A, *inter alia*, expressed its firm conviction that all peoples and all individuals have an inherent right to life, and that the safeguarding of this foremost right is an essential condition for the enjoyment of the entire range of economic, social, and cultural, as well as civil and political rights; stressed the urgent need for all possible efforts by the international community to strengthen peace, remove the threat of war, particularly nuclear war, halt the arms race, and achieve general and complete disarmament under effective international control, and prevent violations of the principles of the Charter of the United Nations regarding the sovereignty and territorial integrity of states and self-determination of peoples.

At the same session, in its resolution 37/189B, the General Assembly, *inter alia*, invited those Member States, specialized agencies and other organizations of the United Nations system that have not yet done so to submit their information in accordance with General Assembly resolution 35/130 A of 11 December 1980, and requested the Commission on Human Rights to give special attention, in its consideration of the item entitled "Human Rights and Scientific and Technological Developments," to the question of the implementation of the provisions of the Declaration.

Document A/38/195, prepared in pursuance of that resolution, contains a report of the Secretary-General based on information received from Member States regarding the implementation of the provisions of the Declaration.

In 1983 and 1984, similar documents (A/39/422 and A/40/493 and Add 1 and 2) were prepared by the Secretary-General to the General Assembly pursuant to resolutions 38/112 and 39/133 respectively.

At its fortieth session, in 1984, in its resolution 1984/27, the Commission on Human Rights invited all Member States and relevant international organizations that had not yet done so to submit to the Secretary-General their views on the most effective ways and means of using the results of scientific and technological developments for the promotion and realization of human rights and fundamental freedoms, and further requested the Sub-Commission on Prevention of Discrimination and Protection of Minorities to consider areas in which studies could be undertaken on the most effective ways and means of using the results of scientific and technological development for the promotion and realization of human rights and fundamental freedoms.

The Commission, in its resolution 1984/28, adopted at the same session, *inter alia* appealed to all states, appropriate organs of the United Nations, specialized agencies and intergovernmental organizations concerned to take the necessary measures to ensure that the results of scientific and technological progress are used exclusively in the interests of international peace, for the benefit of mankind and for promoting and encouraging universal respect for human rights and fundamental freedoms.

Pursuant to its decision 1983/108, by which the Commission on Human Rights decided to consider the item "Human Rights and Scientific and Technological Developments" on a biennial basis, the Commission did not consider that item at its forty-first session, in 1985.

At its forty-second session, after consideration of the question of scientific and technological developments and human rights, the Commission, having expressed its conviction that implementation of the Declaration on the Use of Scientific and Technological Progress in the Interests of Peace and for the Benefit of Mankind by all States would contribute to the strengthening of international peace and security, the economic and social development of peoples, and international co-operation in the field of human rights, stressed the importance, for the promotion of the exercise of human rights and fundamental freedoms under conditions of scientific and technological progress, of the implementation by all states of the provisions and principles contained in the Declaration.

The Commission on Human Rights, in its resolution 1986/10 entitled "Human Rights and Scientific and Technological Developments," also reaffirmed that all peoples and all individuals have an inherent right to life and that the safeguarding of this cardinal right is an essential condition for the enjoyment of the entire range of economic, social, and cultural, as well as civil and political rights and requested the Secretary-General, in the light of the comments and views of Member States, to submit the report on the implementation of this resolution to the Commission at its forty-fourth session in 1988.

Outline of Studies and Reports

During the period between 1971 and 1976, a number of valuable reports on various aspects of human rights and technological developments were prepared by the Secretary-General and the specialized agencies concerned. The major reports of this period were the following:

1. The report on respect for the privacy of individuals and the integrity and sovereignty of nations in the light of advances in recording and other techniques (E/CN.4/1116 and Add 1-3 and Add 3/Corr 1 and Add 4). This report deals with beneficial uses of new devices and methods of auditory and visual surveillance, personality assessment techniques ("personality tests"), polygraphs ("lie detectors"), narco-analysis, certain blood, breath, and other bodily tests for non-medical purposes, and communication and observation satellites.

2. The report on uses of electronics which may affect the rights of the person and the limits which should be placed on such uses in a democratic society (E/CN.4/1142 and Add 1 and 2). It deals with beneficial uses of computerized data systems, benefits derived from computers in various areas of their use, beneficial applications of electronic automation from the point of view of human rights, and benefits derived from electronic communications techniques in various fields of their application.

3. The report on protection of the human personality and its physical and intellectual integrity, in the light of advances in biology, medicine, and biochemistry (E/CN.4/1172 and Corr 1 and Add 1-3). This report deals with the beneficial use of artificial insemination, psychotropic drugs, procedures of pre-natal diagnosis, and chemicals introduced into food production, processing, packaging, and storage.

4. Reports on the impact of scientific and technological developments on economic, social, and cultural rights, which deal with the use in the modern world of scientific and technological progress to increase the availability and improve the quality of food and clothing. These reports also contain information concerning the realization of the rights to food, clothing, just and favourable remuneration, equal pay for equal work, housing, rest and leisure, and social security under conditions of scientific and technological progress (E/CN.4/1084, E/CN.4/1115, E/CN.4/1141 and E/CN.4/1198).

5. The report on protection of broad sectors of the population against social and material inequalities, as well as other harmful effects which may arise from the rise of scientific and technological developments (A/10146). This report deals with the harmful effects of automation and mechanization of production, deterioration of the human environment, the population explosion, and the hazards arising from the increasingly destructive power of modern weapons and from atomic radiation.

6. The report on the balance which should be established between scientific and technological progress and the intellectual, spiritual, cultural, and moral advancement of humanity (E/CN.4/1199 and Add 1). This report deals with the effects of scientific and technological progress upon the enjoyment of particular human rights and fundamental freedoms, including those set out in the following articles of the Universal Declaration of Human Rights: Articles 1, 2, 3, 5, 10, 11(1), 12, 16(1), 17(1), 18, 19, 20(1), 21, 23, 24, 25, 26(1), 26(2), and 27. The report outlines possible international action to assess new technologies, give warning of the possible dangers to human rights which they may present, and possibly even control new developments if threats to human rights seem likely.

7. The United Nations Educational, Scientific, and Cultural Organization's report on the impact of scientific and technological developments on economic, social, and cultural rights, pursuant to Commission resolution 10 (XXVII) of 18 March 1971. It should be noted with interest that the report, E/CN.4/1114, contains chapters not only on the right to culture but also the rights to education and to information. This report is supplemented by document E/CN.4/1196 of 19 November 1975, which contains important considerations on the right to freedom of expression and the campaign against propaganda for war or for national, racial, or religious hatred.

The Commission on Human Rights, at its thirty-third session in 1977, considered a number of reports on various aspects of human rights and scientific and technological developments which it had been unable to examine at earlier sessions, including the report on the protection of the human personality and its physical and intellectual integrity in the light of advances in biology, medicine, and biochemistry,[16] the report on the balance which would be established between scientific and technological progress and the intellectual, spiritual, cultural, and moral advancement of humanity,[17] a report by Unesco dealing with the impact of scientific and technological developments on the rights laid down in Article 26, paragraphs 1 and 2, and Article 27 of the Universal Declaration of Human Rights concerning the right to education, the right to culture, and author's rights,[18] a report on developments relating to science and technology elsewhere in the United Nations system of interest to the Commission,[19] a report on national technological assessment machinery,[20] and a report on the human rights implications of the genetic manipulation of microbes.[21]

On 11 March 1977, the Commission on Human Rights, in resolution 10B (XXXIII), responding to General Assembly resolution 31/128 of 16 December

1976, spoke of the Declaration on the Use of Scientific and Technological Progress in the Interests of Peace and for the Benefit of Mankind as a guide for its future work and re-emphasized the Assembly's concern that Member States should take account of the provisions and principles contained in the Declaration. The Commission's resolution, continuing in this vein, emphasized that Member States should take account of the provisions and principles contained in the Declaration, in particular those relating to the transfer of technology and scientific knowledge to developing countries, which would accelerate the realization of the economic and social rights of the people of those countries. It instructed the Sub-Commission on Prevention of Discrimination and Protection of Minorities to examine, in the light of the provisions of the Declaration, studies relating to this subject, and to submit its observations thereon to the Commission.

Also on 11 March 1977, the Commission on Human Rights in resolution 10A (XXIII) requested the Sub-Commission to study, with a view to formulating guidelines, if possible, the question of the protection of those detained on the grounds of mental ill-health against treatment that might adversely affect the human personality and its physical and intellectual integrity. However, it was not until 1980 that the Sub-Commission appointed a Special Rapporteur to prepare a study on the protection of persons detained on the grounds of mental ill-health.

The Protection of Persons Detained on the Grounds of Mental Ill-Health

The Sub-Commission, at its thirty-third session in 1980, appointed two Special Rapporteurs; one, in resolution 11 (XXXIII) of 10 September 1980, to prepare guidelines relating to procedures for determining whether adequate grounds existed for detaining persons on the grounds of mental ill-health and principles for the protection of persons suffering from mental disorder, and another, in resolution 12 (XXXIII) of 11 September 1980, to undertake a study of guidelines relating to the use of computerized personal files.[22]

The first Special Rapporteur, Mrs Erica-Irene Daes, submitted a preliminary report to the Sub-Commission in 1981[23] and a final report, as requested by the Sub-Commission, in 1982.[24] Having considered the final report of Mrs Daes and having examined the report of the sessional working group on the question of persons detained on the grounds of mental ill-health,[25] the Sub-Commission, by its resolution 1982/34 of 10 September 1982, requested the Commission to recommend to the Economic and Social Council that the Special Rapporteur supplement her final report with an account of the views expressed in the Sub-Commission and any new reply which might be transmitted in the meantime, and also to establish a sessional working group for a proper examination of the body of principles, guidelines, and guarantees and to submit the revised final report to the Commission at its fortieth session.[26] The General Assembly, the Economic amd Social Council, and the Commission on Human Rights subsequently agreed to the above request.

In 1983, at the thirty-sixth session of the Sub-Commission, resolution 1983/39 was adopted which recommended that Mrs Daes' report should be published and given widest possible distribution. The sessional working group continued with its work on the question at the request of the General Assembly, the Commission, and the Economic and Social Council.

In 1984, the Sub-Commission, continuing its first reading of the draft body of principles, decided to consider it further in 1985. In order to expedite consideration of the draft body at the request of the General Assembly, in 1985, the Sub-Commission again established a sessional working group. In fact, the working group approved, at its second preliminary reading, the first seven articles of the draft body of principles, which contained 47 articles. The report of the group was approved, by the Sub-Commission, without a vote. However, no decision was taken on the draft resolution on the question E/CN.4/Sub.2/1985/L.10.[27]

Unfortunately there was no Sub-Commission session in 1986 due to the unforeseeable financial crises of the United Nations. Therefore, no progress was made on the question. In the same year, the General Assembly, in its resolution 41/114, again urged the Sub-Commission to expedite their consideration of the draft body.

In 1987, the working group continued, at its second preliminary reading of the draft body of principles, consideration of articles 8, 9 and 10.[28]

The General Assembly by resolution 42/38 of December 1987 urged expected consideration of the draft body of guidelines, principles and guarantees, in order that the Commission could submit them with its views and recommendations to the Assembly through the Economic and Social Council, it being suggested that such Commission views and recommendations on tne draft should be before the Assembly at its forty-fourth session in 1989. Subsequently, the Commission at its forty-fourth session adopted resolution 1988/62, in which it endorsed the recommendation of the Sub-Commission, in its resolution 1987/22, that the Sub-Commission be requested (a) to attach much greater emphasis at its fortieth session to the Working Group and its drafting assignments; (b) to complete the work on the draft body of guidelines, principles, and guarantees as a matter of urgency at its fortieth session; and (c) to take account of the paper presented by the World Health Organization (E/CN.4/1988/66) and to submit it to the Working Group for consideration.

During the years 1987 and 1988, the World Health Organization and interested non-governmental organizations had met informally to consider the draft body of principles and facilitate agreement between those concerned. In document E/CN.4/1988/66, the World Health Organization proposed that the draft should be in two parts. Part I would include fundamental principles and part II would be detailed guidelines for the implementation of those principles.

In 1988, Professor Claire Palley, the new British independent member, was appointed to chair the Working Group on the Mentally Ill. The other four members of the Group were also new, having just been elected to the Sub-Commission, and therefore to the Group.

Professor Palley from the outset was determined that no efforts should be spared in trying to complete the draft in 1988. She was aware that the four hours alloted to the Working Group during the session of the Sub-Commission would be inadequate for this task. She thus requested that the Group should hold a number of informal consultations. Some 13 hours were spent in meetings in which interested non-governmental organizations and delegates were given the opportunity to make suggestions and comments. During the 1988 session of the Sub-Commission, the Working Group completed consideration of the articles of the draft body of principles, and approved the texts as "Draft Body of Principles and Guarantees for the Protection of Mentally Ill Persons and for the Improvement of Mental Health Care," contained in document E/CH.4/Sub.2/1988/23. The Working Group called this exercise the second preliminary of the draft body

The Sub-Commission subsequently considered the report of the Group which contained the text of the draft body (E/CN.4/Sub.2/1988/23), adopted the draft body, and decided to forward it to the Commission on Human Rights for adoption. The rapid conclusion by the Sub-Commission of this draft was due to the enormous amount of time that Professor Palley devoted to the task, and the new Soviet Expert, Professor Chernichenko, also greatly facilitated completion of the draft.

The Commission on Human Rights, by resolution 1983/40 of 6 March 1989, decided to establish an open-ended working group to examine, revise, and simplify as necessary the draft body of principles and guarantees, and invited comments, for consideration by the working group, from governments and specialized agencies (in particular the World Health Organization and non-governmental organizations) on the draft body of principles and guarantees submitted by the Sub-Commission; it requested the Secretary-General to circulate these comments to governments prior to the session of the working group.

If the Commission and its open-ended working group, after examination of the draft body with comments submitted to it, consider it appropriate to forward it immediately to the General Assembly for final adoption, we might possibly have a new international human rights instrument on the human rights of persons who are mentally ill.

The draft body of principles and guarantees made it clear that all persons with a mental illness shall, in particular, have the right to protection from exploitation, abuse, and degrading treatment. The non-discrimination principle was also confirmed. It was also provided that every mentally ill patient shall have the right, *inter alia*, to receive visitors in private, in particular his lawyer or representative, regularly and at all reasonable times; to send and receive unread and uncensored any private and public communications; and to privacy. It was also significant that a diagnosis of mental illness shall be in accordance with internationally accepted medical standards.

A Study of Guidelines Relating to the Use of Computerized Personal Files

In its resolution 10B (XXXIII) adopted on 11 March 1977, the Commission on Human Rights instructed the Sub-Commission to examine, in the light of the provisions of the Declaration on the Use of Scientific and Technological Progress in the Interests of Peace and for the Benefit of Mankind, studies relating to the implementation of their provisions. As previously mentioned, the Sub-Commission, accordingly, by its resolution 12 (XXXIII) of 11 September 1980, requested its Chairman to designate a member of the Sub-Commission to undertake a study on relevant guidelines in the field of computerized personal files. The Chairman of the Sub-Commission designated Mrs Questiaux for this study.

At the thirty-fourth session of the Sub-Commission in 1981 on behalf of the Special Rapporteur, Mrs Questiaux, her alternate, Mr Louis Joinet, made a statement on the progress report.[29] He stated that the study was not designed to analyse the problem in all its aspects but rather, with due regard for the many studies that had already been made, to examine guidelines which might be taken into account in that field.[30]

At its thirty-fifth session in 1982, this matter was not discussed by the Sub-Commission.

At the thirty-sixth session of the Sub-Commission, the final report was prepared and submitted by Mr Louis Joinet on the relevant guidelines in the field of computerized personal files.[31] At the same session, after approving, in its decision 1983/8, the conclusions and recommendations of the final report on relevant guidelines in the field of computerized personal files presented by Mr Louis Joinet (E/CN.4.Sub.2/1983/18) pursuant to its resolution 12 (XXXIII) of 11 September 1980, the Sub-Commission decided, in accordance with the wish expressed by the Commission on Human Rights in its resolution 10B (XXXIII), to submit Mr Joinet's report to the Commission at its fortieth session for whatever action it deemed appropriate.

At its thirty-seventh session in 1984, the Sub-Commission endorsed the conclusion of the study of the relevant guidelines in the field of computerized personal files prepared by Mr Joinet and requested the Secretary-General to transmit to Member States and all relevant international organizations the provisional guidelines in the field of computerized personal files, inviting them to transmit their views thereon.[32]

At its thirty-eighth session in 1985, the Sub-Commission had before it a revised report by Mr Joinet.[33] Holding the view that, for the purpose of effectiveness, governments should be consulted more widely on the revised draft guidelines, the Sub-Commission requested the Secretary-General to continue to obtain the comments and suggestions of governments on the revised guidelines and to render the Special Rapporteur all the assistance he needed in submitting the final report on guidelines to the Sub-Commission at its fortieth session.[34]

By successive *notes verbales* of 18 November 1985 and 29 April 1987, the Secretary-General accordingly requested the governments to submit comments and suggestions.

At its fortieth session in 1988, the Sub-Commission considered the final report submitted by the Special Rapporteur (E/CN.4/Sub.2/1988/22). The purpose of the report was (1) to identify the main trends emerging from comments made by the members of the Sub-Commission during the discussion of the interim report

(E/CN.4/Sub.2/1985/21), as well as from the analysis of the answers received; and (2) to submit, for the approval of the Sub-Commission, the revised final draft guidelines with a view to their transmission to the Commission on Human Rights. The Special Rapporteur noted that the list of answers received showed the increasing interest of United Nations organs and specialized agencies in the draft guidelines, due, it would seem, to the number of personal files they were keeping.

The Sub-Commission, by resolution 1988/29, welcomed the recommendations contained in the report, in particular the draft guidelines applicable to computerized personal data files, and decided to forward them to the General Assembly through the Commission on Human Rights and the Economic and Social Council. The draft guidelines contain provisions concerning the principle of lawfulness and fairness, the principle of accuracy, the principle of non-discrimination, supervision and penalties, etc. The Special Rapporteur in his report recognized "a consensus [emerging] from the comments received on the desirability of encouraging the formulation of guidelines in this area, both for member states wishing to adopt domestic legislation and for international, organizations and agencies in respect of the status of their own personal files."

In 1989, the Commission, by resolution 1989/43, forwarded the report to the Economic and Social Council.

Further Studies

Pursuant to Commission resolution 1984/27, the Sub-Commission, at its thirty-seventh session, considered areas in which further study would be desirable and adopted resolutions 1984/17 and 1984/18. By these resolutions, the Sub-Commission recommended, through the Commission, to the Economic and Social Council, to entrust the Sub-Commission to undertake a study on the current dimensions and problems arising from unlawful human experimentation and a study on the implications for human rights of recent advances in computer and micro-computer technology.

The Commission on Human Rights, at its forty-first session, determined by its decision 1985/104 to request the Sub-Commission to reconsider the studies mentioned, with a view to integrating them in the work already being undertaken in the Commission and the Sub-Commission under the agenda item "Human Rights and Scientific and Technological Developments."

By its resolution 1984/29, the Commission reiterated its request to the Sub-Commission to undertake as a matter of priority a study on the use of the achievements of scientific and technological progress to ensure the right to work and development.[35]

Another study, which the Commission requested the Sub-Commission to carry out by its resolution 1982/27 of 19 February 1982, concerns the negative consequences of the arms race for the implementation of economic, social, and cultural, as well as civil and political rights, the establishment of the new international economic order, and, above all, the inherent right to life.

At its thirty-eighth session in 1985, the Sub-Commission adopted resolution

1985/7 concerning hazardous processes, products, and technologies. By this resolution, the Sub-Commission, bearing in mind that the effects of scientific and technological development on human rights have both beneficial and harmful aspects, and therefore must be examined in their totality, and recognizing that inadequate information and the absence of uniform protection and safety measures with regard to the potential dangers of the application of hazardous technologies result in a grave threat to the right to health and to life, requested, *inter alia*, all transnational corporations and enterprises to disclose all the information at their disposal regarding the hazards to human lives of their processes, products, and technologies to governments, employees, consumers, and the general public.[36]

The General Assembly, by resolution 42/183 of 11 December 1987, requested the Secretary-General to prepare a comprehensive report on the question of illegal traffic in toxic and dangerous products and wastes for submission to the General Assembly at its forty-forth session in 1989 and to have a preliminary report on the question submitted to the Economic and Social Council at its second regular session of 1988. The report was submitted to the Council as document E/1988/72. In 1988 the Sub-Commission noted that the report drew attention to the increased traffic in toxic and dangerous products and wastes, especially from the developed countries to the developing countries, and recommended that the Commission adopt a resolution (1989/42) on the movement and dumping of toxic and dangerous products and wastes. The Commission subsequently endorsed this recommendation.

OBSERVATIONS ON PAST DEVELOPMENTS IN THE FIELD
AND DIRECTIONS FOR THE FUTURE

Positive and Negative Aspects

We have seen that a great many efforts have been made, including the preparation of a series of studies by various United Nations human rights bodies, to examine both the positive and negative aspects of scientific and technological developments on human rights and to identify the main issues in this field.

Before proceeding further, it might be appropriate to introduce the summary of conclusions drawn from the studies prepared by the Secretary-General between 1973 and 1976, which appear in "Human Rights and Scientific and Technological Developments" of 1982.[37]

The main conclusions that can be drawn from the many aspects of the question of human rights and scientific and technological developments that have been discussed in these chapters are the following:
1. Science being itself a part of culture, the essential problem facing mankind in relation to scientific and technological progress, on the one hand, and the intellectual, spiritual, cultural, and moral advancement of humanity, on the other, is to decide on the appro-

priate two-way relationship which should exist between them. This relationship is not the same for all times or all places.

2. An investigation of this relationship includes an examination of the impact, both beneficial and harmful, of recent scientific and technological developments upon the rights laid down in the Universal Declaration of Human Rights. Such impact affects many such rights, either individually or in combination.

3. The application of policies and measures appropriate to the circumstances is an aspect of achieving the correct relationship between scientific and technological progress and the intellectual, spiritual, cultural, and moral advancement of humanity.

4. Educational policies should aim at a better understanding of science on the part of the general public and a better understanding of the humanities and the needs of society on the part of scientists.

5. Many and varied measures have been taken on the national level for the protection of human rights against threats posed by recent scientific and technological developments. Nevertheless, there is a growing conviction that there is need for continuing technology assessment on the national level in order to assess possible side-effects and long-range effects of new innovations and to establish whether their advantages outweigh the discernible disadvantages, and for control over innovations with harmful potentialities.

6. The positive uses of modern science and technology for the promotion of human rights are potentially vast, but their exploitation depends upon the laying down of appropriate science policies on the national level and the creation of machinery to carry out those policies. Such machinery need not be separate from that set up for technology assessment as envisaged in paragraph 5 above.

7. On the international level, in addition to the possibility of setting international standards relating to specific aspects of human rights and scientific and technological developments, there have been proposals for a general Declaration on Human Rights and Scientific and Technological Developments. There are many existing texts which could be taken into account in drafting such a Declaration.

8. There have also been a number of proposals for technology assessment on the international level. The reasons given for the establishment of machinery for such assessment have been essentially the same as those given in relation to assessment on the national level, except that greater stress has been placed on problems of an international nature, including the prevention of threats to human rights on an international scale and the use of science and technology for the benefit of mankind.

In considering the relationship between scientific and technological developments and human rights, it should not be forgotten that the former does not necessarily constitute a prerequisite for a guarantee of the latter.[38] On this point, the representative of the Federal Republic of Germany submits that "Scientific and technological progress is neither of intrinsic value nor to be evaluated externally, but is to be judged according to the contributions it is able to make for the well being and dignity of the individual, for freedom and the peaceful community of nations."[39]

In other words, science and technology have no limits and they will continue to develop in the future with or without our full-hearted support, and these scientific and technological processes are going to have their effect on human rights.

The main question is how to utilize scientific and technological developments wisely and properly for the benefit of human rights.

In the recent discussion on scientific and technological developments and human rights, it has been pointed out by a number of speakers and observers that the United Nations bodies have mainly been studying the negative nature of any developments and not fully examining their positive aspects.[40] This observation is quite correct as far as the topics of previous studies and resolutions adopted are concerned. However, this observation needs further clarification and explanation.

It is true that the examination, discussion, studies, and research carried out by United Nations human rights bodies should not simply be restricted to analysing the problems; efforts should also be made to seek effective guidelines to secure and safeguard the beneficial usage of scientific and technological developments for the promotion and protection of human rights. However, inevitably in this field, the United Nations had to start examining the issues by identifying the problems at the beginning rather than making efforts to seek guidelines or policy recommendations immediately.

In this connection we should note with interest the statement made by Mr Louis Joinet, when he introduced the progress report on relevant guidelines in the field of computerized personal files, particularly as they affect the privacy of the individual, at the thirty-fourth session of the Sub-Commission.

The study was not designed to analyse the problem in all its aspects, but to examine guidelines which might be taken into account in an international instrument, with due regard for the many studies that had already been made, particularly the Report of the Secretary-General on Human Rights and Scientific and Technological Developments (E/CN.4/1142), which was still relevant and very topical.[41]

The author's interpretation of this statement is that the report by the Secretary-General on Human Rights and Scientific and Technological Developments[42] had already dealt with, *inter alia*, beneficial uses of computerized data systems and benefits derived from computers in various areas of their use, and had basically analysed the phenomenon, and the report presented to the Sub-Commission by Mr Joinet was to focus on how to secure benefits from scientific and technological developments.

Therefore, it is clearly of extreme importance that scientific and technological activities should be guided in such a direction that maximum advantage and minimum disadvantage would result for the effective enjoyment of human rights and fundamental freedoms.

In their examination of the relationship between Human Rights and Scientific and Technological Developments, the General Assembly, the Commission on Human Rights and the Sub-Commission have adopted innumerable resolutions and have prepared a considerable number of studies. It is generally submitted that, so far, most of them fall into the following three categories:

1. Peace, security and disarmament.
2. Transfer of science and technology for effective development of the developing world.
3. So-called "negative aspects of scientific and technological developments," including "computerized personal files and privacy" and "human rights of the mentally ill."

Whilst the author of the present article recognizes the importance of the work undertaken to date, he would also like to emphasize the need for guidelines and rules of conduct or policy recommendations in scientific and technological activities. Or, more precisely, he considers it important that the effective implementation of the existing human rights standards and international law should be recognized in scientific and technological activities and policy-making.

Science and Technology and Human Rights Standards

There is little doubt that scientific and technological developments are the products of scientific and technological activities by human beings. If this is so, it is self-explanatory that those activities should be guided by the existing universally recognized human rights standards and the framework of international law.

Ziman, Sieghard, and Humphrey, in their book entitled *The World of Science and the Rule of Law*,[43] touch upon this point, stating that:

Science is not a disembodied activity, it is a human enterprise, conducted by human beings interacting in a complex social network with each other, and with the societies in which they work. And whatever may have been the case about human rights in the past, today they have acquired a sharply defined and objective framework in which can be set the performance of many human enterprises, including the pursuit of science.

Discussion of the need for guidelines has also been touched upon by the group of experts on the question of "the balance which should be established between scientific and technological progress and the intellectual, spiritual, cultural and moral advancement of humanity."[44] The following quotation is from the conclusions and recommendations section of their report:

3. A thorough revision of education at all levels is required to bring about a sufficient harmony of science and technology with other human activities. Science and technology must be taught in the context of the ascent of Man, not primarily as potential contributors to the disruption of society or the depersonalization of individual lives. A proper understanding of science and its impact on society is essential for dealing adequately with the evolving problems of civilization.
4. Not every change or development that science and technology make feasible needs to become an actuality. Governments and societies must determine by appropriate mechanisms for technological assessment – including the assessment of possible side-effects and long-range effects – whether the time is right for particular innovations and whether their advantages outweigh for the society the discernible disadvantages. International machinery should be entrusted with such a technological assessment for man-

kind as a whole. It is a basic human right to have a voice in such decisions. Decisions in such matters must be made on the basis of the considered opinion of bodies of experts and laymen who represent the interests of all the people as well as of future generations.

5. With these ideas in mind, and taking into account the necessity for keeping under constant review the promotion and protection of human rights in the light of rapid scientific and technological developments, the Group recommends that consideration be given to the possibility of drafting a declaration on human rights and scientific and technological developments.

6. It is recommended that a better definition be given of the duties of the individual to the community and of the right of future generations. For example, it seems to us that the crisis in growth of the world's population must lead to some constraint on the individual right to reproduce, and that the right of the child to be born physically and mentally sound takes precedence over the rights of parents to reproduce.

Concerning this idea of drafting a declaration on human rights and scientific and technological developments, the Federal Republic of Germany states the following:[45]

Concerning the proposal for a declaration on "human rights and scientific and technological developments," the human rights contained in the International Covenant on Civil and Political Rights should first be examined to see whether these do not constitute sufficient guidelines on scientific and technological progress. The field of human rights is characterized more by a lack of implementation than by a lack of legal norms. What is important basically is to improve the instruments for implementation and thereby provide greater protection for human rights.

In this connection, it must be ensured that the institution of "human rights" will not, by the inclusion of global dangers for humanity, be expanded to a point of endangering what has already been attained in the field of human rights protection.

With regard to the creation of a new international institution to deal with questions as to what extent negative influences on human rights protection should be tolerated in order to enjoy the benefits of scientific and technological progress we would refer to existing United Nations bodies (the Commission on Human Rights and the Human Rights Committee). The strengthening and optimal efficiency of these bodies must be the aim of the human rights discussion and activities.

Another recommendation of the group of experts is to arrive at an improved definition of the duties of the individual towards society and the rights of future generations exceeds the scope of human rights.

Whether one agrees or not on the drafting of a new declaration in this field and, if so, on the content of any new declaration, it is quite clear that international standards or guidelines should be borne in mind in this field.

It is evident that not only science and technology but also human rights have a universal transnational character, as it is now true that a more or less complete international code of human rights law exists composed of closely defined rights, freedoms, and duties, agreed upon by the international community of nations, which is independent of any particular place, culture, tradition, or economic or political system, and which has been accepted as an international standard of reference for the exercise of civil, political, economic, social, and cultural activities throughout the world.[46]

It should be recognized that almost all the rights and freedoms of individuals and responsibilities of organs of society and states necessary for the free and effective pursuit and activities of science and technology are in fact already covered by the existing international code of human rights law. This code actually covers, in principle, all men and women, including scientists, beneficiaries, and users of science and technology. Consequently, if the code could be fully implemented, positive aspects of science and technological developments would be more effectively safeguarded and promoted.

In conclusion, if we want to utilize science and technology for the benefit of mankind (in other words, "human rights"), we have to respect and observe human rights standards at the same time.

In this context, the representative of Yugoslavia made an interesting intervention:[47]

The idea of using the results of scientific and technological developments for the promotion of human rights is part of the broader efforts to place material development in the service of man. In this, emphasis should be placed on the promotion of human dignity, man's position in the society and the elimination of his dependence and alienation, together with the improvement of his general living conditions.

The link between scientific and technological developments and human rights, in particular those which have been unequivocally recognized at the international level, to whose observance numerous states have subscribed, can be examined proceeding, first of all, from the relevant provisions of the International Covenant on Economic, Social, and Cultural Rights as well as the International Covenant on Civil and Political Rights.

The right to education (Article 13) and the right to take part in cultural life (Article 15) are in particular closely linked to the progress in science and technology. Article 15, paragraph 1, recognizes the right of everyone to enjoy the benefits of scientific progress. Paragraph 2 of the same article stipulates that steps to be taken by States Parties to the Covenant to achieve the full realization of this right shall include those necessary for the conservation, the development and the diffusion of science and culture. Paragraph 4 emphasizes the importance of international co-operation in the scientific and cultural fields for the realization of these rights.

However, it would be erroneous to concentrate in the current review only on economic, social, and cultural rights. Civil and political rights are not, as it is often claimed, only endangered by scientific and technological progress but they frequently to a large extent depend on them. Certain rights stipulated by the International Covenant on Civil and Political Rights can well illustrate this notion.

The realization of the right to freedom of expression (Article 19), including the freedom to seek, receive and impart information and ideas cannot be imagined today without a developed system of mass media communication and without ensuring to a large number of individuals and groups access to these media.

Development of science and technology, coupled with an adequate social policy, can play a significant role in the ensuring of material conditions for genuine enjoyment of the formally proclaimed and adopted civil and political rights. Science and technology should also be used in the improvement of the organization and operation of institutions for the protection of these rights, by making them more easily accessible to larger portions of the poorer populations.

Modern Science and Technology and Freedom of Information

The development of science and technology is opening up new horizons of knowledge and is making possible the acquisition of previously undreamed-of vast sources of energy, the conquest of outer space and the exploitation of the depths and sources of the seven seas. The scientific and technological revolution underway offers mankind powerful means for solving global problems, eliminating hunger, disease, poverty and illiteracy, overcoming economic backwardness.[48]

Indeed, the development of science and technology in the modern world is remarkably fast and constant. On an everyday basis, new inventions and discoveries are made in every corner of the world.

Modern science and technology is, however, so sophisticated and complex that only people who have already studied the subject, and not the layman, are likely to understand the complicated implications of specific scientific or technological developments. The whole process of science and technology is most of the time no longer just a matter of doing simple experiments with a test tube and bunsen burner. The process by which new scientific and technological ideas are generated and tested might take many years, and test results might also be interpreted differently. In these circumstances, it is very difficult to know or to foretell what would lead to positive results and what to negative results, nor what sort of developments would be regarded as unlawful or unjustifiable.

In addition to these difficulties, more impediments exist in modern society of concern to individuals controlling the direction of scientific and technological developments. Now, many of the major scientific and technological activities are carried out by state-related agencies or big commercial industries in secrecy for national or industrial security reasons. All over the world, scientists are most of the time dependent on their governments or big industries. And, of course, this dependence can expose scientists to some explicit or implicit pressures not to disclose information.[49]

One writer elaborates this point with regard to hazardous exports.[50] She says that "if the end-user, whether called consumer or simply a human being, has not been alerted to the hazards which imported goods, foreign in the strict sense of the word, carry with them, there is no manner in which he/she can defend him/herself. The person becomes a defenceless victim, defenceless because of being kept ignorant of the danger imminent in the goods given, or bought. . . ."

There is a need for everyone to defend himself or herself against the negative effects of science and technology; to do so one needs access to information on the potential dangers of specific scientific and technological developments, be it pesticide or baby food, nuclear power plants or pharmaceuticals, as negative effects are experienced in different parts of the world.

Therefore, it is freedom of information that is at the heart of the protection and promotion of all other human rights. The European Commission on Human Rights has asserted the principle of free flow of information to the public in general.

It is only after obtaining the correct information on a specific question that people can act with some degree of confidence, perhaps participating in a decision-making process or appealing before different legal, quasi-legal, administrative, or legislative bodies. These legal and procedural aspects are largely covered by civil and political rights. Indeed, these civil and political aspects are important for the effective realization of economic, social, and cultural rights. This "interdependent nature of economic, social and cultural rights and civil and political rights" is quite obvious in this context.

Concerning the freedom of information and free flow of information for the beneficial use of scientific and technological developments, it is interesting to note the efforts of the United Nations Educational, Scientific, and Cultural Organization[51] and the International Social Science Council, a non-governmental organization,[52] in this field.[53]

In March 1985, the International Social Science Council held an international symposium at Barcelona to consider the challenges to human rights posed by scientific and technological progress. Scholars in the field of anthropology, biochemistry, biology, brain research, clinical, experimental and physiological psychology, genetics, law, mental health, neuroscience, pharmacology, and psychiatry attended the symposium and contributed papers on a variety of scientific subjects, with particular reference to their impact on human rights. Special attention was paid to the human rights set forth in Articles 5 (the right to be free from torture or cruel, inhuman, or degrading treatment), 26 (the right to education), and 27 (the right to enjoy cultural life) of the Universal Declaration of Human Rights; in Articles 12(1) (the right to enjoy the highest attainable standard of physical and mental health), 13(1) (the right to education) and 15(1) (the rights to take part in cultural life, to enjoy the benefits of scientific progress and its applications, and to benefit from the protection of the moral and material interests resulting from any scientific, literary, or artistic production) of the International Covenant on Economic, Social and Cultural Rights; and in Article 7 (the right to be free from torture or cruel, inhuman, or degrading treatment) of the International Covenant on Civil and Political Rights.

The participants at the symposium agreed that it was important that Unesco should educate the general public not only regarding the application of scientific discoveries for the benefit of human rights, but also on the potential danger to human rights of withholding or abuse of scientific discoveries. It was recognized that that would only be possible if the general public had a sound basic education in the behavioural and social sciences, which would allow it to appreciate the benefits of science.

Consequently, it was decided to create a network of scientists concerned with human rights issues, who would exchange information on ways of promoting human rights in their research activities.

The symposium made five priority recommendations to Unesco, that it should:
1. Seek the broadest possible worldwide applications of tested scientific advances and promote scientific contributions to human rights.
2. Encourage scientists to become aware of human rights issues.
3. Pay special attention to the impact of new technologies on the human rights of women, particularly with respect to the productive processes.
4. Give special priority to the human rights of patients, a particularly vulnerable group.
5. Develop standards for the use of neuropathetic drugs.

It further recommended that experts in neuroscience and the behavioural disciplines should be consulted when determining policies and administering programmes in education and health.

CONCLUDING REMARKS

The Universal Declaration of Human Rights, adopted in 1948, must be counted as one of the great achievements of the United Nations. The Declaration and the Covenants and more than 50 other conventions that grew out of it provided the world, for the first time, with an international code of human rights which establishes as norms of international law the way in which the state must treat individuals. Specific mechanisms have also been established by the United Nations to monitor compliance with these agreements.[54]

Nevertheless, this significant and historical change in this field has not deterred scientists and engineers, politicians, businessmen, philosophers, and citizens from continuing to debate whether or not scientific and technological activities and decision-making processes would be bound even by a moral obligation.[55] It may come as a surprise to some that scientific and technological activities and decision-making processes have something to do with international human rights standards.[56] These current policies, approaches, and attitudes which create or maintain an artificial dichotomy between legal and human rights considerations and scientific and technological activities should be abandoned or rectified. Imperative human rights, humanitarian and legal considerations are too often considered as separate and secondary interests, or sometimes even regarded as irrelevant and not as integral parts of scientific and technological activities and policies. One might add that this approach of "artificial compartmentalization" could be for many policy- and decision-makers a convenient means to marginalize the human rights and legal factors.[57]

In this connection, the comments made by the United Nations Department of Technical Co-operation for Development, in its reply to the questionnaire on "popular participation"[58] pursuant to resolution 1986/14 of the Commission on Human Rights, should be noted with interest: "The notion of human rights underlies technical assistance activities, and more specifically, UNDTCD proj-

ects in the field of integrated rural development, concerned with the improvement of socio-economic and living conditions in rural areas, based on the concept
of community and popular participation in all aspects of decision-making, as well
as in the implementation of practical activities affecting their lives."

Participation in the field of politics, culture, education, communication, social
science, etc., is not an end in itself. It also serves as a means for the exercise of
human rights in these spheres, at the local, national, regional, and international
level.[59] Professor Peter Jambrek pointed out in his study, entitled *Participation as a
Human Right and as a Means of Exercise of Human Rights*, prepared for the Unesco
Division of Human Rights and Peace in 1982, that "Participation is also viewed as
a means to guarantee and foster the right to development, to combat racism and
racial discrimination, and to promote the interests and needs of underprivileged
communities in the process of decision-making at political, economic, social, and
cultural levels."[60]

It has been ascertained that the United Nations University has developed a
series of projects on scientific and technological developments, and it has been
explained that these projects have the common objective of setting forth the
guidelines for science and technology policies to contribute best to human and
social development needs.[61]

In fact, the need for guidelines has been pointed out by the group of experts on
the question of "the balance which should be established between scientific and
technological progress and the intellectual, spiritual cultural, and moral advancement of humanity"[62] as well as by the summary of conclusions drawn from the
studies by the Secretary-General between 1973 and 1976 in "Human Rights and
Scientific and Technological Developments."[63]

Whether one considers or not that the existing human rights standards already
cover the whole range of scientific and technological activities, or that some additional standards are necessary, it could be said that it is now time that we recognized the need to establish the accountability of governments and of international
organizations in terms of the impact of their policies and programmes on the
beneficial use of scientific and technological developments from a human rights
and international legal point of view in the present world.

In 1984, the head of the Japanese delegation to the Commission on Human
Rights, Mrs Sadako Ogata, stated that: "the time had come to appraise the studies
undertaken and bring them up to date where necessary in the light of the advances
made during the past decade, so as to enable the Commission to consider what
further action was needed."[64]

Asserting that "the existing human rights standards in the field are adequate so
that no further guidelines would be necessary" may not necessarily be persuasive
as an argument, when we know that effective implementation of the existing
norms is something different from mere standard-setting. It might be interesting
to recall, in this connection, that the Commission on Human Rights, although
recognizing that new standards, rules, or new rights are not necessary, decided to
prepare a new "declaration on the right and responsibility of individuals, groups

and organs of society to promote and protect universally recognized human rights and fundamental freedoms" by establishing an open-ended pre-sessional working group of the Commission.[65] This future declaration's main aim is not to establish new rights but to promote more effective implementation of the existing human rights standards. A similar approach could be also taken with regard to science and technology.

It is submitted that establishing new guidelines in the field of scientific and technological developments and a new human rights framework might certainly facilitate the transfer of scientific and technological activities from the purely technical and natural scientific arena to the legal and political arena, so that more human rights considerations would be given in the process.

The author of the present article truly hopes that the United Nations University's project on scientific and technological developments and human rights will bring us some new and fruitful achievements in the field, particularly as this is the first time the United Nations University has decided to respond to an invitation made by a United Nations human rights body and to launch a project of this kind.[66]

NOTES

1. The original version of this paper was published in *Nordic Journal of international Law*, vol. 56, no. 4 (1987), as "Consideration of the Relationship between Scientific and Technological Developments and Human Rights in the United Nations Human Rights Fora."
2. Adopted at the 50th meeting, on 10 March 1986, without a vote. The text of the resolution appears in document E/1986/22.
3. E/CN.4/1986/SR.50, para. 63.
4. E/CN.4/1986/SR.21, paras. 18–22.
5. E/CN.4/1986/SR.22, paras. 6–7.
6. E/1986/22.
7. Project document dated 20 November 1986, UNU.
8. Project document (note 7 above), "2. Objective."
9. E/CN.4/1986/28.
10. Horst Keilau, "Scientific and Technological Progress and Humanism," in Daniel Premont et al, eds., *Essays on the Concept of a "Right to Life" in Memory of Yougindra Khushalani* (Bruyland, Brussels, 1987).
11. The International Conference on Human Rights was held in Tehran, Iran, from 22 April to 13 May 1968 in connection with the observance of the International Year for Human Rights, which marked the twentieth anniversary of the Universal Declaration of Human Rights. The purposes of the conference were to review the progress made in the field of human rights since the adoption of the Declaration, to evaluate the effectiveness of the methods used by the United Nations in the human rights field, and to formulate and prepare a programme of further measures to be taken subsequent to the celebration of the anniversary. The governments of 84 states were represented at the conference, which adopted the Proclamation of Tehran and 29 resolutions. These

resolutions, and the general principles set out in the Proclamation, gave new directions to a great deal of the subsequent work of the United Nations in the field of human rights. See United Nations, *Final Act of the International Conference of Human Rights* (UN Sales no. E.68.XIV.2).

12. EN/CN.4/1023 and Add 1–3 and Add 3/Corr 1 and 2 and Add 4–6.
13. For a summary of the discussion on the report, see E/4949, paras. 175–192; E/CN.4/SR 1119–1127.
14. A/9937, para. 11.
15. A/10330.
16. E/CN.4/1172 and Corr 1 and Add 1–3.
17. E/CN.4/1199 and Add 1.
18. E/CN.4/1196.
19. E/CN.4/1234.
20. E/CN.4/1235.
21. E/CN.4/1236.
22. See below.
23. E/CN.4/Sub.2/474.
24. E/CN.4/Sub.2/1982/16.
25. E/CN.4/Sub.2/1982/17.
26. E/CN.4/Sub.2/1982/SR.36 and E/CN.4/1982/4, chap. 1, section A, draft resolution IX.
27. Sub-Commission decision 1985/112. See E/CN.4/1986/5, paras. 317–323.
28. E/CN.4/Sub.2/1987/32.
29. E/CN.4/Sub.2/490.
30. E/CN.4/1512, paras. 205–209.
31. E/CN.4/Sub.2/1983/18.
32. Sub-Commission resolution 1984/12.
33. E/CN.4/Sub.2/1985/21.
34. Sub-Commission resolution 1985/14.
35. The Commission had made such a request in 1981 and 1982 in resolutions 38 (XXXVII) of 12 March 1981 and 1982/4 of 9 February 1982.
36. E/CN.4/1986/5, paras. 307–312.
37. United Nations, *Human Rights and Scientific and Technological Developments*, pp. 91–92 (DPI/726-41527-1982-DM).
38. E/CN.4/1986/SR.21, paras. 19–20.
39. E/CN.4/1986/28,
40. For example, E/CN.4/1986/SR.21, paras. 15, 16, 21, and 22.
41. E/CN.4/1512, para. 206.
42. E/CN.4/1142.
43. J.M. Ziman, P. Sieghard, and J. Humphrey, *The World of Science and The Rule of Law* (Oxford University Press, 1986).
44. E/CN.4/1199 of 2 February 1974, para. 4.
45. Ziman et al. (note 43 above), p. 6.
46. Ziman et al. (note 43 above), p. 6.
47. Intervention at 1983 Commission on Human Rights.
48. Reply of the Union of Soviet Socialist Republics of 13 February 1985, in E/CN.4/1986/28.
49. Ziman et al. (note 43 above).

50. K. Tomasevski, "Hazardous Exports: Access to Information Necessary for Self-protection," paper prepared for the 10th University of Windsor Symposium on Law and Development Combating Process of Rural Impoverishment: Law as an Obstacle and a Resource for Victims, 5–8 November 1986, p. 1.
51. See also *United Nations Action in the Field of Human Rights*, p. 256 (ST/HR/Rev.2), (UN Sales no. E.83XIV.2).
52. A non-governmental organization in consultative status with the Economic and Social Council (category II).
53. E/CN.4/1986/SR.21.
54. J. Pérez de Cuellar, *Report of the Secretary-General on the Work of the Organization* (1985), p. 17.
55. P. Alston, "International Law and the Human Right to Food," in P. Alston and K. Tomasevski, eds., *The Right to Food* (Nijhoff, The Hague, 1984), p. 12.
56. Regarding the question of law and human rights in the wake of technology, see C.G. Weeramantry, *The Slumbering Sentinels* (Penguin Books, 1983).
57. Theo van Boven, "Book Review on Alston and Tomasevski, The Right to Food," *Netherlands International Law Review*, vol. 3, no. 1 (1986).
58. E/CN.4/1987/11.
59. P. Jambrek, "Participation as a Human Right and as a Means for Exercise of Human Rights" (Paris, 1982), p. 1 (Unesco doc. SS-82/WS/54).
60. Jambrek (note 59 above).
61. Project document (note 7 above).
62. See note 44 above.
63. United Nations (note 37 above).
64. E/CN.4/1984/SR.28, para. 11.
65. E/CN.4/1986/40 and E/CN.4/1987/38.
66. Preliminary Report on the Project on Human Rights and Scientific and Technological Development was submitted by the United Nations University to the Commission on Human Rights at its forty-fourth session, E/CN.4/1987/48. Subsequently, the Commission adopted resolution 1988/59, in which it invited the United Nations University, in co-operation with other academic and research institutions, to continue to study both the positive and the negative impact of scientific and technological developments on human rights and fundamental freedoms and expressed the hope that the United Nations University would inform the Commission at its forty-sixth session of the results of its study on the question.

BIBLIOGRAPHY

Daes, E.-I. *Principles, Guidelines and Guarantees for the Protection of Persons Detained on Grounds of Mental Ill-Health or Suffering from Mental Disorder.* (UN Sales no. E.85 XIV 9; UN doc. E/CN.4/Sub.2/1983/17/Rev.1.)

Jambrek, P. Participation as a Human Right and as a Means for Exercise of Human Rights. Unesco, Paris, 1982. (Unesco doc. 55–82/WS/54.)

Keilau, Horst. Scientific and Technological Progress and Humanism. In: Daniel Premont et al., ed., *Essays on the Concept of a "Right to Life" in memory of Yougindra Khushalani.* Bruyland, Brussels, 1987.

Kubota, Yo. Consideration of the Relationship between Scientific and Technological Developments and Human Rights in the United Nations Human Rights Fora. *Nordic Journal of International Law*, vol. 56, no. 4 (1987).

UNCTAD. An International Code of Conduct on Transfer of Technology. (UNCTAD doc. TD/B/C.6/AC.1/2/Suppl.1./Rev.1,1975.)

United Nations. *Human Rights and Scientific and Technological Developments*. (DPI/726-41527-1982-DM.)

Weeramantry, C.G. *The Slumbering Sentinels*. Penguin Books, 1983.

Ziman, J.M., P. Sieghard, and J. Humphrey. *The World of Science and the Role of Law*. Oxford University Press, 1986.

Part 4

Some Specific Issues

7

Human Rights and the Structure of the Scientific Enterprise

SHIGERU NAKAYAMA

INTRODUCTION

Scientists open up many problems of human rights and lawyers attempt to solve them. So far, however, the two communities have remained isolated from each other at two extremities of a spectrum.

In a search for bibliographical information on the matter of "human rights and science," I made some inquiries of my "science study" community, which includes the disciplines of history, philosophy, sociology of science, and science policy studies. The typical responses of my colleagues have been that our study has reached the sociological, economic, political and even ethical aspects of science, but not the legal one as yet.

Though there have been numerous writings on human rights issues, it seems that the issue of human rights in relation to the development of science and technology has been raised only recently, and that people have developed a "critical awareness of science"[1] only since the late 1960s. The turning-point may be found in the "Human Rights and Scientific and Technological Developments" resolution at the Intergovernmental Conference on Human Rights held at Tehran in May 1968.

Since then, a number of possible sources of violation of human dignity have emerged, along with the rapid development of the frontiers of micro-electronics and life science, as enumerated by Professor C.G. Weeramantry in his recent book *The Slumbering Sentinels*, which is, to my knowledge, the first serious attempt to tackle the issues from a legal point of view.[2] The tempo has been so rapid that the legal profession seems not to have caught up effectively. This state of affairs has been lamented by my lawyer friends who have no scientific background. Weeramantry too has pointed out that the law has been tardy in evolving concepts to deal with technology.

Scientists in their laboratories, preoccupied with their immediate goals, are not aware of the possible consequences of their research. Legalists deal with the issue of science and technology in an *ex post facto* fashion on such aspects as society and

Table 1

	Academic	Industrial	Defence	Service
Assessors	Peer	Sponsor	Military	Public
Competition	Individual	Corporate	National	None
Expression	Open	Classified	Classified	Open

environment, aiming to fill the gap between the scientific and legal communities. I shall explore the production mechanisms of scientific knowledge and technological information in the human rights context, and then present the general pattern of attempts to discover where principles and mechanisms for technology assessment can best be applied. This sort of analysis is, in my view, more essential than the piecemeal information gathered on the research front.

CLASSIFICATION OF SCIENCE BY ASSESSORS

As suggested by Dr Mushakoji in his problématique for this project, science and technology should be treated as information. There are two ends of information flow: the production of information at one end and its assessment and utilization at the other. In the post-war period, the latter has come to be overwhelmingly powerful while the former tends to occupy a subordinate position. The central concern here is who will review and assess the information that scientists and technologists produce. From the viewpoint of the sociology of science, I hold the view that the question of the audience to whom scientific research is addressed is the most important factor in shaping and defining its character. Viewed in terms of operative mechanisms, four types of scientific activities may be distinguished, as shown in table 1.

From this point on, I do not in principle distinguish between science and technology, as this is only a historically meaningful distinction, which may no longer be valid, particularly in the case of industrial and defence sciences (though the classical distinction may still hold good at times in the case of academic science).

STRUCTURE OF ACADEMIC SCIENCE

Academic science is the science practised in the open scientific community, in which members present their research for debate and discussion by their colleagues, and seek recognition of their work through publication in scientific journals. Research is initiated out of personal interest and pursued for reasons of personal honour and distinction meted out through a referee system.

In academic science, researchers typically proceed by the following steps:
1. Design of research.
2. Application for research funds.

3. Research.
4. Peer review of research findings.
5. Publication.

In the above process, steps 1 and 3 are matters of highly individual concern, since it is generally believed that classical freedom of research has to be maintained. This is the point at which academic science becomes very different from industrial and defence sciences.

In reality, however, it has been proven that nineteenth-century science can no longer compete with post-war organized mass science and hence, as the size of scientific enterprises inevitably grows bigger, it becomes difficult to maintain individual identity in formulating research problématiques, and the tendency is to conform to the norms of mass science. Sometimes the human rights of non-conforming scientists and engineers are infringed so that the immediate goal of a bureaucratic organization may be achieved, as in the case of the promotion of the SDI project. Though not always overly careful, researchers usually keep their ideas and findings secret before publication, since they are concerned about the possibility of their being stolen by competitors.

In the post-war period, practically all scientists have applied for research funds of some kind. All academic scientists are involved in the competition for securing research funds, success in which process often predetermines the winner of the game at an early stage, even before any research is commenced. At this stage the distinction between members of the scientific community and amateurs is sharply applied, placing the latter out of reach of any research funds. This situation contributes to the formation of scientific groups resembling closed-shop unions, resulting in a clear-cut demarcation between the scientific community and those outside it. In the scientific community application forms are commonly sent to an assessment committee and are subject to close scrutiny by peers. It is also not uncommon for applicants preparing their forms to conceal carefully the essential points of their ideas, so that a peer assessor cannot appropriate them.

In the 1970s, when the unforeseeable dangers of DNA were recognized, it was felt that the social evaluation of science should not be *ex post facto*, on the outcome of research, but preliminary, before the research starts. The National Institute of Health tried to use their funding mechanism for the pre-assessment of hazardous research by applying their guidelines. Since corporations could afford such research with their own funds, this standard could not be extended to industrial science.

At step 3, a general statement is hardly possible because of the diversity of topics involved in scientific endeavour. One recent trend is observable, however.

Unlike a social scientist, who clings to ideology and conscience in carrying out research, an experimental scientist is happy to change his methodology until he finally arrives at a satisfactory position. In a major scientific project, however, he cannot easily do so owing to the fear that he will be called upon to account for funds already spent, even though he might know that the project will eventually turn out to be a failure, in such cases as nuclear fusion and the SDI project.

At step 4 another opportunity for "post-research and pre-publication" assessment is theoretically open. Again, on manuscripts submitted to the referee mechanism, precautions are taken to defend the researchers' prior claims to the knowledge they contain. Otherwise, a referee might hold up its publication and meanwhile appropriate the ideas and publish them elsewhere.

Along with overspecialization in science and the overproduction of research output, it is generally suspected that the old referee system is approaching bankruptcy, though people still believe that distinguished work can survive the process of selection.

In view of the recent advent of micro-electronic technology, step 5 deserves to be treated in the following independent subsection.

INEFFECTUAL ACADEMIC FREEDOM OF EXPRESSION

In the old days, knowledge and information were the private monopoly of a select group of people, transmitted orally or secretly through manuscripts. The modern printing media have made knowledge publicly available and promoted the objectivity of modern science and freedom of expression for the last three centuries. With the referee system intervening between the manuscript and printing stages, science was raised to the status of public knowledge.[3]

In reality, however, academic science is defenceless in safeguarding human rights against authority, ecclesiastical or civil.

Freedom of expression is still "academically" maintained, though the knowledge and information produced are circulated within a closed circuit, practically out of the reach of most people. Furthermore, the new micro-electronic media will bypass the referee system that modern scientific tradition has so far cultivated and turn scientific knowledge back into privately circulated and monopolized information.[4] In the generality of cases, micro-electronic media have been commercially developed to promote private information rather than public knowledge. There is a fear that by making knowledge private once again we will revive the exclusiveness of knowledge and information which was a feature of the manuscript age.

Moreover, as the new electronic media make it possible to disseminate small quantities of a wide range of scientific information, the trend towards specialization will be spurred on still further, beyond the reach and comprehension of the general public.

GENERAL REMARKS ON INDUSTRIAL AND DEFENCE SCIENCES

Industrial science as practised in private laboratories has had much in common with defence science and little in common with academic science. Jerry Ravetz generalized its many manifestations into a single term, "industrialized science,"[5] emphasizing the characteristics of production of scientific information as similar to those of an industrial commodity. David Dixson would like to call them "strategic science," with emphasis on its goal-oriented nature.[6] I prefer the Japanese term *taiseika kagaku* (Establishment Science) which stresses its characteristic of tight and rigid incorporation into the present establishment and its isolation from the general public; thus, it may be called, with Steven Rose, "incorporated science."[7]

Although most people still talk of science in terms of the classical academic science paradigm, this is not the dominant form of contemporary research. Major resources for research and development are now allocated to industrial and defence sciences. These incorporated sciences are vigorously promoted by the driving force of competition among corporations or nations, while profits and military strategic necessity determine how their results will be evaluated.

Industrial science is targeted to the areas that promise the greatest profits. Defence science proceeds along policy lines laid down by a handful of military strategists, despite the fact that it involves a large amount of taxpayers' money. The most problematic aspect of these sciences is that they continue to remain beyond the purview of public scrutiny or any form of public accountability. The researcher's work is assessed wholly within the organizations, and is not reviewed or evaluated by anyone from outside. This is why industrial and defence sciences are referred to more generally as "incorporated sciences."

Scientific works openly assessed and publicly recognized are called "knowledge," while the findings of incorporated sciences, classified and available only to a limited circle for private monopoly, are referred to as "information."

THE STRUCTURE OF INDUSTRIAL SCIENCE

The sociology of science as an established discipline has developed considerably during the last two decades, but so far its major interest has continued to be the analysis of traditional academic science. We badly need the structural analysis of incorporated science but, mainly because of its very corporate nature, we have to confess that the investigation is neither thorough nor penetrating as yet. Furthermore, there is no general rule applicable to all establishments that are competing with each other. However, it is still worthwhile to conceptualize the idealized case.

The steps of commercialized industrial science may be summed up as follows:
1. Market demand.
2. Targeting and formulation of plans.
3. Research and development.
4. Model, test, and production.
5. Advertisement.
6. Dissemination into market.

In step 1, consumer assessment is incorporated when business enterprises carry out market research. A leader in each research group is responsible for formulating his or her target at step 2, taking into account the overriding need for commercial success, and submits his or her research proposal to the upper strata of the corporation. Unlike academic science, researchers in industrial science cannot determine their targets autonomously and hence their creativity is said to be considerably reduced.

At step 3, the assessment by top managers of corporations comes in. Many research plans are abandoned at this stage. In industrial science, assessment by top managers and sponsors, rather than the wishes of consumers, tends to prevail in determining the final product at step 6.

Steps 2 to 4 are absolute corporate secrets. Confidentiality is maintained against competing companies, not against consuming customers, but its ultimate effect appears to be similar.

Old Marxists claimed that the large corporation, with its desire to maintain market stability, purchased the patents of scientists and inventors and kept them secret, resulting in the distortion and negation of healthy progress in science and technology; in other words, monopoly capitalism killed off competition in science and technology. There have been many cases in history to prove the above statement, but the present-day reality is that industrial science is ruthlessly promoted by competition between rival enterprises in the oligopolistic system.

With regard to research and development, it is cynically remarked in the scientific community that, in rapidly changing and innovative industrial science, the findings of primary importance are classified as the "know-how" of the company and those of secondary importance are patented and sold on the market, while only insignificant results from the corporation's viewpoint are reported in scientific journals and academic meetings following the convention of academic science. In the most innovative fields, however, intellectual property rights cannot be maintained too long as such property becomes outdated quite quickly.

At the step of basic research, classification is not yet too strict, but as the production step is approached, precaution is exercised to an excessive degree. Social assessment can be applied only to steps 5 and 6. Step 1 is then returned to, thus repeating the cycle. It is, however, often pointed out that post-advertisement assessment is too late, especially on the frontier of biotechnology and life science, where past experience and simulation do not help much.

STRUCTURE OF DEFENCE SCIENCE

The structure of defence or armaments science is the most difficult for us to analyse, as it is deeply embedded in the self-perpetuating military-industrial complex. The following model is taken from the well-publicized, recently formulated SDI Project of the USA.

1. Idea and design.
2. Proposal to the military.
3. Governmental approval and parliamentary hearing.

4. Research and development.
5. Model, test, procurement, and production.
6. Assessment by the military.

Whereas all the steps in industrial science are normally conducted within the corporate system, the common procedure in respect of defence science projects is twofold: scientists in universities and corporations initiate ideas and military procurers assess and accept them. In reality, the research sectors of all the above organizations form an exclusive group called a military–industrial–academic complex. Hence, whether research is conducted in universities, corporations, or in-house military laboratories, such an activity can be called defence science.

The market demand for industrial science noted in step 1 is absent in the case of defence science in peacetime. Ideas for new weapons development have never originated from professional career servicemen but have been initiated by scientists who have made proposals to the military; the SDI project is such a case. The military always seek quantitative expansion while qualitative leaps are made only by scientists.[8]

The professional assignment of scientists and technologists is to create something new. Thus, researchers in a military–industrial–academic complex constantly feel compelled to propose ideas for new weaponry. On the other hand, the military establishment can no longer assess the cost of innovation of weapons systems at step 2. Matters are entirely left in the hands of a handful of leading scientists and technologists.

At this point, a word is in order to clarify the nature of the "military–industrial–academic complex." This does not mean the weapons production departments of corporations but the complex of research and development sectors of universities, national laboratories, corporations, and military establishments. In the case of the SDI project, the Lawrence-Levermore Laboratory, MIT, and the Lockheed Corporation are major sharers of the governmental R&D budget, where step 1 to step 4 are proceeded with and the results fed back to step 1 in a repetitive cycle. It would not necessarily be continued to step 5 of procurement and production. The complex is only happy as long as the cycle continues to be repeated and escalated. This is a point where defence or incorporated science differs from industrial science. Like a nuclear fusion project, incorporated science can exist by implementing a cyclic mechanism without reference to immediate production.

From the point of view of the social assessment of science, the problem here is obviously the exclusiveness of the project, its confinement within a very limited circle. Step 3, the only step available for social assessment, is usually avoided in the name of strategic secrecy. In step 3, the process of governmental approval and parliamentary hearing on the R&D budget proposals, public assessment is formally carried out, but in practice certain limitations on public inquiry are imposed even at the congressional hearing, under the pretext of national security. In place of market demand, the most compelling arguments are built up on the assumed effectiveness of the weapons system of a hypothetical enemy.

Research and development in step 4, in the case of defence and incorporated science, is conducted on a large scale, on the solid base of a huge national budget with which industrial science is unable to compete. The high social status of defence scientists and technologists in the American scientific and engineering community is due to the huge budget with which they can develop their ideas freely, without being disturbed by commercial market assessments.

Sometimes a rationale is provided for a high defence science budget on the basis that its findings will have a spin-off to the civil sector, thus enabling industrial science to enjoy economies in the initial cost of R&D, which private corporations cannot afford. A little thought would show these arguments to be incorrect, as defence science has to observe, except in step 2, a strictly enforced classification which prevents its findings from leaking to other social sectors. Only in those areas such as nuclear energy and space science, where the basic paradigms of research have originated in defence science but have public applicability, have spin-offs been possible. But these areas could not be called industrial science; they are, rather, incorporated science, or pseudo-civil science, as Aant Elzinga calls it.

At step 5 the position of defence science is similar to that of industrial science, but the lack of commercial assessment often leads to extravagance; the organization may have proven its efficiency for the immediate purpose of wartime crash programmes, but those scientists involved, if they work on a long enough project, can lose a sense of economy and can often be misled to corruption, which is still more prevalent in step 6.

In the final stages of step 6, peacetime assessment of new weaponry is made at practice manoeuvres, but the difference from real war is obvious. Weapons systems can best be assessed only in terms of the difference between them and those of opponents. Assessment becomes a game within the military-industrial-academic complex against a hypothetical enemy. In the last step, a new military demand and requests for further improvement will appear, as necessary for defence against a hypothetical enemy, resulting in the renewal of the cycle.

HUMAN RIGHTS OF SCIENTISTS AND ENGINEERS

Scientists' rights to publish research results are often infringed or denied because of the policy of giving priority to the interests of the business and military establishments. Especially for those scientists and engineers in industrial and defence sciences, these rights are categorically denied. As the Unesco Recommendation on the Position of Scientific Researchers (1974) has already demanded, the right to disclose military or corporate secrets, in cases where scientific information is of a crucial nature for human existence, should be protected.

I cannot, however, be too optimistic about the ombudsman function of the scientific community. My long experience shows that only a small portion of the community (say 10 per cent) is truly qualified as guardians of human rights, the rest being prone to undertake whatever research is best funded. Most contemporary scientists are involved with the industrial and defence establishments, and thus the collective inclination of the scientific community is not necessarily sound.

Important Role of Technicians

I would like to suggest in this connection the important role to be played by technicians, or rank and file scientists and engineers, rather than leaders of projects, in affording protection against science-related hazards.

In general terms, there are three stages in the development of pollution originating from scientific research.

1. Small-scale pioneering experiment by a leading scientist.
2. Large scale R&D in major industrial laboratories.
3. Industrialization and introduction of industrial products into the environment.

In stage 1 the scale of experiment is so small that its hazards are faced only by the individual scientist who designs the project. It is well known that Madame Marie Curie shortened her natural life by exposing herself to radiation at her laboratory. Early workers on the DNA recombinant experiment must also have undergone such risks. In such cases, as well as those of pioneer-adventurers, intellectual or otherwise, an individual risk may be compensated by individual success and reputation. As long as the adventurer maintains complete freedom of decision-making in exposing his life to the hazards of his experiment, no infringement of his human rights will occur. Furthermore, in the early stage the rule of the openness of academic science is usually observed.

In stage 2 large-scale R&D involves a number of technicians and rank and file scientists who are engaged in the project for their livelihood. Their risk may be compensated partially by their salary, but this is not worth the risk they are exposed to. Hence, they can assume a more cautious and detached attitude than the adventurous leader in identifying hazards. If they have no freedom to reveal and express themselves about possible hazards, there will have been an infringement of their human rights.

In the case of X-ray experiments, the International Physics Society admitted and set a threshold value of X-ray exposure only as late as 1928, when many technicians were exposed to the radiation hazard. In the case of the recent recombinant DNA experiment, labour unions such as the British Association of Scientific Workers have played a watch-dog role in formulating guidelines for the experiment through assessment committees like the G-MAG (Genetic Manipulation Advice Group). The Association of Scientific Workers has long been concerned to protect the human rights of technicians and laboratory workers.

These hazards are a sort of occupational disease that may turn into environmental pollution when scaled up.

In stage 3 the citizenry has nothing by way of compensation. They may retain the right to complain about pollution, but they have no special knowledge of it and hence they can easily be cheated in the ensuing debates. In the case of the recombinant DNA experiment, a representative of the citizenry was invited to the G–MAG. Jerry Ravetz, a critical historian of science, was chosen for that work but he complained that his role was merely ornamental. It was then said that G–MAG was a cosmetic exercise and Jerry Ravetz was the lipstick. None of the scientists on G–MAG believed there was a real hazard.

It is rather difficult to find an expert who stands on the side of the citizenry on hazards and pollution issues. At this point, it is indispensably important for the technicians at stage 2, with their expert knowledge, to disclose and expose to the public the possible hazard before it can be diffused into the environment.

STRUCTURE OF SERVICE SCIENCE

The slogan "science for mankind" still lingers in the popular imagination. Scientists who devote themselves to "truth-seeking" are considered to be qualified to decide what is truth and what is not, to be providers of intellectual services to mankind and, like other professionals, to be quite independent of earthly desire for wealth and power. A classic example would be those engaged in the search for bacteria at the turn of the century.

I have defined "service science" as science assessed by the citizenry – citizenry defined as "those who have no direct vested interest in science and technology activities as such" and who are thus qualified to be objective, disinterested assessors. It seems that there is no such assessing mechanism by the citizenry at work at the present time, but again a classic exception may be found in medical science, in the relationship between a doctor and his client.

As we cannot expect much from academic science as a counterbalance to the menace of incorporated industrial and defence science, we must inevitably turn to another kind of enlightenment, that of service science.

For the role of protecting human rights against the aggression of industrial and defence sciences, service science needs to be promoted in place of classical academic science. Indeed, we need to elaborate the concept of service science a little further and to develop the strategy and tactics for promoting it.

Its structure is rather simple, since nothing, no bureaucracy, no interest group, intervenes between practising scientists and citizen assessors; it is as follows:
1. Exposure of problems.
2. Solution of problems.
3. Assessment by the people.
All of the above processes are, in principle, kept open to the general public.

In step 1 we find outspoken messages of service science expressed by science journalism, which exposes all positive and negative aspects of scientific endeavour to the people for their own critical judgment and choice. The incident of the Asilomar Conference in 1974 marked the beginning of science journalism's assumption of this critical role in a positive sense.

Journalists present serious problems but never solve them. Journalistic provocations are relatively short-lived. Before society tires of repeated alarms, some-one must take a step toward solving the problem.

It is the official duty of the non-military sector of public laboratories to under-take such service science. The daily life of taxpayers is conducted not on the basis of competition but rather compassion and co-operation, and so also should ser-vice science be conducted for the betterment of ordinary life. However, in the absence of such competition as has existed in the academic and incorporated sci-ences, service science for the people's sake lacks adequate financial support and motivation and remains insignificant and powerless.

In point of fact, if there is any one element that distinguishes service science from any other type of science, it is the system by which research is reviewed and evaluated. Service science addresses its findings neither to fellow scholars, as in academic science, nor to research administrators, as in incorporated science, but directly to local residents. Yet, service science still does not have any clearly de-fined apparatus for this purpose. In reality, works of scientists at a service science institute, like a research institute for environmental protection, are still evaluated according to the standards of academic science.

Hence, service science requires forms of communication that differ from the scientific papers of academic science or the know-how and reports of incorpo-rated science. Since neither originality nor the accumulation of classified know-how is involved in service science, it can make its point in handbills and appeals, in a style designed to secure as wide an audience as possible. It also makes max-imum use of journalism and the broadcast media.[9]

NEW ASPECTS OF SERVICE SCIENCES

So far, we have discussed service science in a positive way as a defence against the hazards of incorporated sciences. Medical and life sciences used to be primarily service sciences of the first grade and there was little intervention from industry or the military between scientists and the general public.

The service science of life is, however, increasingly industrialized now. Medic-al technology assessment is needed in connection with the introduction of elabo-rate and expensive machinery into the health-care system. Mass production of artificial organs will be the big industry of tomorrow, far surpassing the car in-dustry of today.

The advent of the new life sciences portends future markets for the sale and hire of human organs and sale of human tissues. We do not know as yet how much these businesses will be industrialized. As they are still conducted between an individual doctor and an individual patient, people may not be aware of the possi-ble danger of human rights infringement. In spite of the direct influence of life science on everyday life, laymen are quite at a loss in evaluating and assessing the outcomes of research.

RIGHTS OF THE IGNORANT

It is obvious that the human right of access to scientific information should be fully guaranteed, in principle. Towards this end we shall introduce a new issue hitherto not much considered nor discussed. We would like to bring to your attention an issue that perhaps makes up two sides of the same coin, namely "the human right not to know about specific scientific information but still not to remain at a disadvantage because of ignorance" or, in brief, "rights of the ignorant."

It may be too early to propose such a right without fear of being misunderstood, especially when enlightened education still needs to be promoted among the people in the third world. But we can foresee that it is a logical consequence of contemporary trends that the infringement of human rights, due to the compartmentalization of scientific and technological knowledge, is destined to become serious sooner or later. This is by no means an official proposal being formally presented, but we would like to provoke our readers to consider this issue.

Suppose that maldistribution of information is modified to the extent that nobody suffers any disadvantages from inaccessibility to information resources. Still a fundamental problem remains unsolved. It is a deep-rooted issue in the contemporary intellectual environment. It is an ever-present issue in relations between technocratic specialists and the amateur citizenry.

Suppose that the technocratic establishment starts to construct an atomic power station in a region. Local residents naturally feel apprehensive of possible radioactive pollution. The technocrats then hold meetings to discuss devices to dispel residents' misgivings. They start with the premise that knowledge of all sorts concerning atomic power is held by the establishment and that residents are entirely ignorant of that knowledge. Their next move is to send out pamphlets to residents illustrating the safe nature of atomic power or to send specialists to a public forum to explain it to the residents.

The method of persuasion on the part of the scientific establishment may be as follows: the distrust of or misgivings about an atomic reactor on the part of the citizenry is due to their ignorance on scientific matters and hence they must be required to be knowledgeable in such matters to the same standard as experts. If they learn enough, they will be convinced of the safety of atomic power. It is entirely because of their lack of effort in gaining information about science and technology that they oppose governmental planning.

Resident groups may not be convinced by the persuasive arguments of the establishment, and may sometimes invite critics who are public activists in order to have a counter view.

These residents are in the main people who live average lives and have only a basic knowledge of reading, writing, and arithmetic. Then, a new body of knowledge about atomic power is forced upon them. Unless they work hard to gain knowledge about atomic power and build up their ability to assess the incoming new technology, they may not be able to survive in the atomic age. Hence, whether they like it or not, they will have to devote much effort to understanding recent advances in atomic power technology, and even basic atomic physics, in order to prevent themselves being deceived. From the viewpoint of the citizenry, this is nothing but a pollution of the intellectual environment, or information pollution; namely, laymen are requested to be knowledgeable in the technicalities of atomic reactors, which is a nuisance to those who do not have any particular training in atomic science. They would like to spend their time more enjoyably. Further, the kind of technology with which citizens are incessantly called upon to familiarize themselves is not a technology of quality.

In these circumstances, how much claim can the citizenry make to the human right to remain ignorant about any particular knowledge?

In the debate on environmental issues, techno-bureaucrats often insist that they are not concealing information (say, of possible hazards of atomic reactors) from the citizenry, but the citizenry complains that, as they do not have time to specialize in atomic physics, they cannot effectively assess what the technocrat-specialists are trying to bring into their local environment. This gap of information, or more precisely this gap in the time allocated to specialized knowledge by specialists and ordinary citizens, will continue to widen so long as scientific specialization keeps advancing.

In order to prevent this situation of "no information, but retained advantage," we shall have to introduce a human right "not to be at a disadvantage even though one does not have training and time for mastering specialized information." A similar problem may be detected in arguments over bilingual education in large American cities. Do Hispanic high-school students have a right to be illiterate in English? Unless such a human right is introduced, future society will be full of information pollution, from which nobody will be able to escape.

NOTES

1. Jerry Ravetz's term.
2. C.G. Weeramantry, *The Slumbering Sentinels* (Penguin, 1984), pp. 17–21.
3. John Ziman, *Public Knowledge* (Cambridge University Press, 1968).
4. Shigeru Nakayama, "The Three Stage Development of Knowledge and the Media," *Historia scientiarum*, no. 31 (1986): 101–113.
5. Jerome R. Ravetz, *Scientific Knowledge and its Social Problems* (Oxford University Press, 1971).
6. Private communication, November 1987, Paris.
7. Shigeru Nakayama, *Shimin notameno kagakuron.* (Science Studies for the Citizenry) (Shakai Hyouronsha, Tokyo, 1984).
8. S. Zuckerman, *Nuclear Illusion and Reality* (Viking Press, New York, 1982), pp. 103, 145.
9. Shigeru Nakayama, "The Future of Research – A Call for a Service Science," *Fundamenta scientiae*, vol. 2, no. 1 (1981): 85–97.

BIBLIOGRAPHY

Morris-Suzuki, Tessa. *Beyond Computopia: Information, Automation and Democracy in Japan.* Kegan Paul International, 1988.
Nakayama, Shigeru. The Future of Research – A Call for a Service Science. *Fundamenta scientiae*, vol. 2, no. 1 (1981): 85–97.
——. The Three Stage Development of Knowledge and the Media. *Historia scientiarum*, no. 31 (1986): 101–113.
Ravetz, Jerome R. *Scientific Knowledge and its Social Problems.* Oxford University Press, 1971.
Weeramantry, C.G. *Nuclear Weapons and Scientific Responsibility.* Longwood, 1987.

8

Human Rights, Technology, and Development

C.G. WEERAMANTRY

In 1986 the General Assembly in its Declaration on the Right to Development (GA Res 41/128 of 4 December 1986) formulated the right to development in terms that "The right to development is an inalienable human right by virtue of which every human person and all peoples are entitled to participate in, contribute to, and enjoy economic, social, cultural, and political development, in which all human rights and fundamental freedoms can be fully realized."

It is not proposed in this chapter to enter into jurisprudential discussions on the question of the status of this right in terms of international law or rigorous legal analysis. It is sufficient for our purposes if we recognize its existence, even as an evolving norm.

Nor is it proposed to attempt definitions of the concept of development, except to observe that development will not be taken as being confined to economic or material advancement, but is conceived of rather as "the upward movement of the entire social system and in conditions which afford individual members of the society the opportunity to benefit personally from the upward movement."[1]

To what extent, then, can technology (I use the term here as a shorthand expression for "science and technology," especially as it is the practical applications of science that concern us most in relation to the developing world) be used to promote the objectives encapsulated in the General Assembly's definition of the right to development?

It will be seen that there are three principal elements involved, namely
(1) participation in decision-making regarding the introduction of a new technology,
(2) contribution to the creation of the technology in question; and
(3) enjoyment of the development resulting from technology.
The development itself could be economic, social, cultural, or political.

PARTICIPATION

Where technology is of a sophisticated nature the stages of participation and contribution tend to be long delayed. There is also a greater tendency for the creators
and the owners of the technology to keep to themselves the decision-making
process regarding their product and adaptations. They tend to delay even more
resolutely the process of scientific input into the technology itself. Thus, far from
there being a spontaneous spread of technology and industrialization as some
theorists once expected (cf. theories of the spontaneous spread of capitalism), the
events especially of the past half century have shown that technology which is of
monetary value tends to become concentrated in the hands of those who create or
own it, thus confining it largely to the developed world. This is despite theories
of freedom of scientific knowledge. Although there is a desperate need for that
technology in the developing world, the latter seems unable to draw the technology to itself and absorb it into its own social and cultural milieu. If it comes in, it
does so with strings attached which can be manipulated from afar. It assumes a
place in developing societies very much like that of a foreign body in a biological
system, which never integrates fully with its host.

Contrary therefore to the theory that knowledge is universal, technological
knowledge seems to belong and to retain the semblance of belonging to the developed world, wherever it may be actually applied.

A principal reason for this is the lack of third-world participation and contribution, as outlined above. Through such lack of participation, attitudes have
evolved of viewing modern technology as a distant and foreign thing.

This sentiment often manifests itself in a plan for the total rejection of Western
science, such as was voiced by some delegates at the recent conference on "Crisis
in Modern Science" held at Penang, Malaysia in 1987 and hosted by the Third
World Network and the Consumers' Association of Penang. Some advocates of
Islamic and Hindu science, pointing to the undoubted fact that science had made
considerable independent and pioneering progress under these systems, sought to
detach science from its existing international base and build it up anew on their
own cultural base.[2]

This is in this writer's view not a realistic option and the universalism of science
must be adopted even in countries which see themselves as its exploited victims.
It is by participation and contribution that such attitudes of alienation can be
averted and full use made of the beneficial potential of science and technology for
every society. Everything traditional can sometimes be romanticized at too high a
cost. An effort is therefore required to accelerate the ability of developing countries to participate in and contribute to such technology.

This brings us to an issue which lies very much at the heart of the problem – the
way in which it tends to be assumed that the developing world's role is that of
passive receiver of technology rather than of active determinant of the particular
technology it will receive. It is true there are many situations where the freedom

to choose is circumscribed, but at the same time the range of choices available is often much larger than is commonly supposed.

Elsewhere in this volume, and especially in the chapter by Dr Chamarik, the point is made that a conscious choice on the part of the recipient country must determine what technology it will accept or reject. That choice is not purely a scientific one, but depends upon social and economic factors and the expertise of many disciplines. This becomes all the more important when we realize that science is essentially a social product. Its development in Europe reflected a particular socio-economic context and as it developed it interacted in that socio-economic context to provide a new set of needs which it began to serve. Technology has an even more intimate interaction in society than pure science.[3] All of this interaction, heavily influenced by Western needs and norms, can well render a particular technology ill suited for a totally different third-world context – hence the importance of participation in determining the technology.

A prime factor in this process of determination is the principle that technological choice should turn in the direction of technologies which liberate the many rather than facilitate control of the few.[4] Unfortunately the choice of technology in the developing world is often in the hands of those who would like it to be an instrument in the furtherance of their own power – the power of the controlling élite. Unfortunately also, there are many technologies, particularly in the information area, which facilitate processes of social and political control.

The bending of technology to service rather than oppression necessitates a preliminary examination of the routes through which a new technology finds its way into developing countries. We shall therefore look at some ways in which the developing world's participation in the decision-making process can be strengthened.

Routes of Entry

A new technology finds its way into a developing country either:
(1) by the country's own decision, uninfluenced by external factors;
(2) under the pressure of external factors which for one reason or another it is unable to resist; or
(3) through a combination of internal and external factors.

It is the author's view, which needs to be empirically tested, that the technologies introduced under head (2) are more numerous and extensive than is commonly acknowledged and that they tend to be the dominant technologies in nearly every developing country.

I will deal briefly with each of these in turn.

1. Internal Decisions Uninfluenced by External Factors

A developing country may take a decision to import a new technology because its government

(a) genuinely believes that that technology is truly essential to the well-being or development of the country;

(b) believes that there will be a political advantage to it in importing such technology, e.g. that computerization of the governmental process will give it greater control over its citizens or that some extensive project, e.g. an irrigation project, will gain it political mileage;

(c) is under pressure from internal vested interests to import such technology. Such pressures may come from economic interests it needs to cultivate for political reasons or from powerful individuals who will develop a vested interest in the new technology;

(d) has corrupt reasons, in the sense that individual members of the government or their associates may see private profit in it for themselves;

(e) sees advantages in the offer of a technology highly attractive to the dominant classes in the target country, though not directly of benefit to the masses. Such an élitist class and its demands are often psychologically linked to those of the consumer society of the West.[5] As Marc Nerfin, the President of the International Foundation for Development Alternatives, so cogently observed at a UNU Workshop in Indonesia in 1986,[6] "there is a fracturing into two of every society, much worse than the traditional East–West and North–South rifts: the two Indias, the two Chiles, the two USAs, the two worlds."

The list is not exhaustive. It becomes important to gather information regarding the factors preceding the decisional process. If there are circumstances pointing to the existence of such factors as are listed in (b), (c) and (d) above, they need analysis and exposure. This is not always easy, for we are here on sensitive ground. It is better, however, that the reality be recognized and be taken into consideration than that we should continue paying obeisance to the myth that the problem disappears when the decision is a purely autonomous one.

2. Pressure of External Factors

The government may be powerless to resist a new technology or product for many reasons, e.g.:

(a) It may be tied to an aid package and must therefore be taken along with it.

(b) It may be a technology sought to be introduced by a multinational corporation which is so influential in the country concerned that the government cannot afford to displease it and has no option but to permit its introduction.

(c) It may be associated with a military or economic alliance or grouping and the country concerned is hence not in a position to resist it.

(d) The country concerned may have adopted an open-door economic policy and feels it may damage its image as a free investment country by denying entry to a technology which is heavily pushed by powerful overseas financial interests.

(e) The technology emanates from a country which is the most powerful trading partner of the developing country concerned. Any refusal to permit the importation of the technology could have severely damaging effects upon the trading relationship, to the grave detriment of the country concerned.

(f) The technology emerges from a country to which the country concerned is heavily in debt. Similar factors operate as in (e) above.

(g) The technology may be one associated with the arms trade. The armaments industry, already in a relationship with the country concerned, presses it and the country cannot resist.

(h) Massive advertising campaigns launched with all the powerful economic resources of the developed world may create a captive market in the target country, which exercises compelling pressures upon the government to permit admission of the project.

(i) The technology concerned may be made available apparently free of charge, but the strings attached are that the providing country becomes the sole market for the apparatus that needs to be purchased for implementing the technology, e.g. television infrastructure or a high-technology hospital.

3. Mixed Internal and External Factors

There could be a blend of one or more of the factors enumerated above, in numerous permutations and combinations.

The decisional process prior to the creation or importation of the new technology must take acount of all these possibilities, each in its own way a very real factor finally resulting in the entry of unsuitable or oppressive technologies. The factors outlined above have serious implications not only in regard to the prevention of harmful technology but also as inhibiting factors preventing access to technology which could foster and promote human rights, as well as free choice of such technologies. Compulsion to choose a given technology in a limited economic situation often means compulsion to forgo another more beneficial or more desired technology.

Indeed, the matter does not end here, but these influences continue after the introduction of the technology. It brings more unsuitable technologies in its train as the original technology, once entrenched, becomes too expensive to dislodge.

It is vital to the connection between technology and the right to development that there be a clearer understanding of these decisional processes and of the complex of factors that contribute to them. This is an important area of analysis in any study. It will be a useful research project to analyse in the context of two or three countries the decisional routes of entry into them of perhaps five selected technologies.

Such enhanced understanding is an important means of introducing more real participation in the decision-making that introduces technology into developing countries. Often these factors convert an apparent freedom to decide into a situation of non-freedom – a situation where the vast bulk of a developing country's population has the decision thrust upon them that they will receive an unsuitable or exploitative technology.

Such considerations lead logically to the need for the more widespread diffusion of information in those communities, regarding not only the alternative technologies available but also the inhibiting decisional factors outlined above. This involves an educational process which will produce results in the long term, but in the short term it necessitates the setting up of multidisciplinary committees charged with the task of assessing the new technologies in the light of the country's economic, social, and cultural needs. Purely scientific committees are inadequate to cope with these decisions, as has now been well accepted for several years.[7] A greater awareness of the technological alternatives is important to the work of these groups.

In other words, there has to be a break from the deterministic theory regarding technology which holds that technology runs an inevitable course and that it cannot be resisted as it has a motive force of its own.

Technology Surveillance

In scanning the means by which the developing world's participation in decision-making can be strengthened, the concept of technology assessment assumes great importance. Technology has a way of taking over the control of areas to which it applies. Once a sophisticated technology is introduced – be it in the field of information, engineering or agriculture – it tends to dictate its own requirements and make all else subservient to its own needs. Often the very problem itself, which the technology was intended to serve, becomes subordinate to the servicing of the technology. The intended servant becomes the master.

It is necessary, especially in developing societies, to keep constantly in view the principle that technology is there to serve the people and not to be served by them. There should be control over it at every point. This means among other things a continual process of review by committees or groups, of which citizens form an important element. The experts are no doubt necessary, but there needs to be constant questioning as to whether the technology is serving its purpose, creating new problems or being in fact counter-productive from the standpoint of the needs of any particular country.

A matter to be considered is whether in relation to developing countries mechanisms ought not to be recommended which will achieve this purpose.

Just as developing countries all too often exhibit an attitude of resignation to the inevitability of technology, they also entertain a feeling of futility in regard to any attempt to control it. It is both too complex and too powerful to admit of lay attempts at control. That attitude needs re-examination.[8]

It may well set trends in many developing countries if this study broadly recommends an essential human rights policy related to technology, suggesting that institutional structures for the constant surveillance of technology be set up as a matter of course. The habit of technology surveillance does not exist in most developing societies and it must be stimulated to emerge.

The techniques of technology assessment, well developed and entrenched in the affluent world, can afford useful guidelines. We have here an ironic situation. New technologies are evolved largely in the developed world and are hence suited for the developed world. They are scrutinized in the developed world by technological assessment committees. Yet before their introduction to the developing world, for which they are far more likely to be unsuitable, they are not scrutinized by comparable boards or committees. Indeed, the need for such scrutiny in the developing world is even greater, for another reason. The developed world has already felt the impact of most of these technologies and adjusted to it, while the developing world has yet to feel the first impact of many of them. That impact must be assessed before the technology is foisted on them, for it can disrupt many factors essential to the stability of those societies; one example might be a labour-saving technology such as robotization for which that country is not yet ready.

A large segment of the industrial production in the third world is directly or indirectly affected by new technologies – e.g. microchips and the creation and use of new materials. Even seemingly remote technologies such as organ transplants can assume human rights dimensions in the developing world, as for example when they generate a traffic in organ donation between the rich and the poor worlds. Vigilance is required at all points of possible contact between new technologies and the developing world.

Often a correct choice cannot be made for lack of knowledge of the new technologies on the part of local decision-makers. It is essential therefore that decision-makers in third-world situations be better informed of the entire concept of technology assessment.

We cannot, of course, leave this topic without referring to the Bhopal disaster and the surveillance systems resulting from it. The chemical hazard control programme launched by India is a very ambitious one. It includes identification, analysis, and control of all industrial activities involving potentially hazardous chemicals and processes. It also involves a census of India's estimated 5,000 chemical production units and a wide range of safety measures, including a greatly strengthened factory inspectorate. Unless developing countries think in these terms they are more than likely to be dumping grounds for careless or untested technology, with consequent damage to human rights.[9] If, on the other hand, they are prepared to think in these terms it will make possible a great enhancement of the enjoyment of technology.

Early Recognition and Alert Systems

When a new technology is introduced there should be early recognition and alert systems in regard to its impact. With this end in view, workshops have already been organized such as the International Workshop on Advanced Technology Alert Systems: Towards Exchange of Experiences and Promotion of International Co-operation in Technology Assessment, which was organized by the German Foundation for International Development (DSE), in West Berlin in December

1985. The deliberations of that committee brought out the fact that the assessment of technology linked with technology management and technology transfer policy is indispensable. Professor Rohatgi from India, one of the participants in this workshop, declared: "We feel this application of formal and quantitative techniques of technology forecasting assessment and alert systems could lead to a vastly improved and comprehensive identification of opportunities and emerging technologies for developing countries well in advance."

Private corporations in India have recently begun to conduct forecasting studies relevant to their own research and production codes. The general use of such methods by administrators is, however, still far away.

The Advanced Technology Alert System (ATAS) offered by the United Nations Committee on Science and Technology for Development is also making a substantial contribution to the early recognition of possible applications of new technologies. One of its earlier studies was on biotechnology and microchips. However, the output is not as steady and continuous as is necessary, considering the vast nature of the problem and the numerous countries that need this expertise. Other international organizations like ILO and UNIDO offer international monitoring and warning systems, especially in the fields of microelectronics, genetic engineering, biotechnology, and materials. One of the objectives that can be pursued through the current programme is a concentration upon the ways in which knowledge of technology assessment can be diffused throughout the developing world. It would indeed be true to say that the new technologies can, before their full impact is sufficiently realized, make a great dent in the cultural and social progress of third-world countries unless thought is given in advance to their impact upon the social systems concerned.

Pursuing these notions further, technology assessment *prior* to the creation of the product can lead to the fashioning of a product more suitable to the recipient's needs and social and economic conditions. This is an important feature of the rights of protection against the harmful effects of scientific and technological developments. It is a vital part of the informational process preceding the evolution of an appropriate product. It is also an important way in which developing societies can make a greater *contribution* to development in the area of technology, as suggested in the General Assembly's Resolution on the right to development.

International Exchanges of Technological Assessment

A network of information exchange between quality control and standards bureaux of all countries needs to be established in a co-operative spirit. This committee can perhaps contribute substantially to the evolution of such a network by studying ways and means for the collaboration and exchange of such information and by making facilities available for the exchange of policing and investigating services, information and rules.

The Dag Hammarskjöld Seminar for Alternative Development strategies in the Southern African region, 1985, suggested a regional weekly newsletter linked to local, district, and national papers for the exchange of technological information pertinent to the region. Such newsletters do not involve the expense involved in full third-world circulation and need to be considered.[10]

Technical Co-operation among Developing Countries

In the midst of all these discussions we must not lose sight of the fact that there are great technical resources within the developing world itself. Much of this has evolved with special reference to the needs of the developing world. It is true that it comprises only a small percentage of the totality of global scientific and technological resources, but it is nevertheless significant enough for a concerted effort to be made to share this around among the developing countries themselves. They can do this with little expense to themselves, and the efforts that have so far been made for technical co-operation amongst developing countries (TCDC) need to be stepped up. TCDC is an idea which has already produced considerable results. The problem relates also to the question of information flow, which we shall discuss later. There needs to be a network of communication among scientists of the developing world, and it could well be that the United Nations University can contribute to the global effort to promote such an exchange of information as well as of personnel.

Development Digest already provides a valuable information link for this purpose, and perhaps its linkage with district and national papers can be explored.

Such a network of information is also evolving in the form of ASSET (Abstracts of Selected Solar Energy Technology), and the work that has been done in the ASSET field could well be duplicated in other fields. Attention needs to be given to linking together technologists and scientists in particular areas of expertise from various developing countries.

Fields in which such exchanges of information and experience are obviously required are solar energy, tropical health and diseases, mining, health, agricultural procedures, low-cost housing, and cottage industries, to mention but a few. All of these are areas where technology can well be exchanged among third-world countries. It would be helpful if the journals in question also carried a list of experts in those fields, along with their addresses.

The programme of action on the establishment of a new international economic order states that "developing countries are urged to promote and establish effective instruments of co-operation in the fields of science and technology, transport, shipping and mass communication media." This can only be brought about if proper vehicles of communication can be devised. One of the means towards this end is that of an exchange of scientists and technologists; an exchange fellowship scheme needs perhaps to be considered in this context.

It is worth recalling that it was as long ago as December 1972 that the UN General Assembly passed a Resolution which authorized UNDP to convene a special working group to examine ways for developing countries "to share their capacities and experiences with one another with a view to increasing and improving development assistance"; one of the several factors which were identified as tending to inhibit technical co-operation among developing countries was the lack of communication and information systems in relation to the capacities and requirements of developing countries. There has been insufficient progress along this route in the decade and a half that has elapsed since then. There was much discussion at the time regarding an "attitudinal barrier" which favoured the use of experts, consultant firms, and equipment from developed countries. Efforts must be made to overcome that attitudinal barrier.

With more participation by developing countries they will have a greater ability to resist deterministic beliefs about the inevitability of technological change. As an Australian study advocated for Australia, itself often a receiver of technology, "Technological change is not inevitable and predetermined. We must resist deterministic talk about inexorable change. . . There is and must be choice. We can, and must, influence choices made in technological decision-making."[11]

CONTRIBUTION

We have dealt thus far with participation in decision-making about the introduction of new technology. If developing-world scientists and decision-makers are involved at the stage of evolution of a new technology, there can be an even more substantial attempt to turn technology to the service of development.

As a writer on the subject has observed:[12]

Techniques do not exist in heaven, in Platonic caves or in entrepreneurs' imagination, ready to be plucked from the air and incorporated into use. Techniques, whether they be methods of administration or machines to produce consumer goods, have to be invented, developed, introduced, modified, etc. The development of techniques is essentially a historical process in which one technique with one set of characteristics replaces another in the light of the historical and economic circumstances of the time. The historical nature of technological development means that the time and circumstances in which any particular technique is developed heavily influence its characteristics.

At the same time we must be conscious of practical restraints, such as that a multinational can rely on expertise from all over the world to develop a project, while a local operator, who comes to the project for the first time, will have to make all the mistakes the multinational has already made and still not possess a fraction of the latter's experience.[13]

We shall look at some facets of these problems.

Participation in Product Designing

The needs of the country concerned, together with the dangers and the advantages, should wherever possible be part of the input into the *designing* of the technology. This avoids the unsatisfactory situation of the developing country being presented with a completed technology or technological product which it must either accept or reject. Every country has its own requirements and is entitled to products tailored to those requirements. This may not always be feasible but in many cases where it is in fact feasible the unsuitable product tends to be designed without adequate participation from the recipient country.

An important guideline for the export of technology should be that where a particular developing-world market is sought, representatives from the recipient country should wherever possible be brought into the designing stage prior to the final completion of the product. It is true that the forces of the market-place sometimes bring about that result, but the formulation of such guidelines will assist in making this a routine procedure wherever possible.

Joint Venture Enterprises

Where a technology is involved which is too sophisticated for the developing country to produce by itself, joint venture enterprises would be essential. The developing world would have a substantial share in these and hence be able to contribute its own expertise to the process of production.

Research in international trade law aimed at giving more strength to the local component of joint venture enterprises can establish more ways in which that local component can assert itself. At present the foreign participant often enjoys dominance by virtue of being the owner of the technology.

Research would need to cover the basic transfer modes of technology, such as licensing or franchising, subcontracting, and supply of equipment, materials, or technology. Much technology comes in the form of total package transfer, including such matters as market survey, product mix, design, production plan, quality control, and manufacturing processes and procedures. Also involved are the foreign exchange component and managing and marketing skills.[14] Selected aspects of this package can be scrutinized with a view to increasing the local input. Indeed, there should be the capability of untying the whole package in order to scrutinize its different component elements with a view to maximizing the local input in consonance with local needs. Tied packages of conditions which must be taken or rejected as a whole often enable very exploitative conditions to be foisted upon recipients. There is an important field of work here for the lawyers and international consultants in the developing world. Currently the legal expertise available in most parts of the developing world cannot match the expertise that can be commanded by the joint venture partners of the developed world. Perhaps the United Nations University or the Human Rights Commission could assist in this respect by organizing seminars or training courses for developing world lawyers and administrators in this field.

Where there is a process of subcontracting, the relationship between the large or parent enterprise and the subcontractor needs to move away from the "parent–child" type of relationship to one of equal partnership. Local legislation can provide additional strengths to the local partner, making the bargaining relationship closer to one of equality.

Where the transfer is through a supply of equipment, material, and basic technology there will be considerable room for adaptation to local needs. This will also generate technological skills in the recipient countries.

In every one of these types of operation the importance of the generation of local skills cannot be overemphasized.

An important area of legal research involved in all of these types of technology transfer is in the area of restrictive trade practices. Inherent in many of them are restrictions which severely inhibit local input and the spread of local expertise. For the socio-economic development of a nation the latter is particularly important and cannot be left to the chance forces of the market-place. It requires deliberate policy measures by the government of the recipient country. Indeed, government agencies, such as the Council of Science and Industrial Research set up by the Government of India, or Institutes of Small-scale Industries or National Science Development Boards, can help in co-ordination and in giving advice on some of the technical aspects involved in such policies. Non-governmental organizations, such as trade associations and industrial co-operatives, can also provide these services.[15]

The process of technology transfer to developing countries often comes up against a problem sometimes described as "the double economic structure." By this is meant the fact that in many of these countries there is, in and around the large cities, a prosperous and sophisticated consumer group, often with Western-oriented tastes, while in the other areas there is a large population with different background and tastes. Both are potential groups of customers. To which groups does one cater?[16]

There are heavy human rights overtones in the decision to be made. The chances are that the economically and often politically more powerful urban group would be the target of the introduced technology. The decision-makers invariably come from this group. Their needs tend to be supplied while the needs of the rest, however urgent, pass unheeded. Sophisticated cookers for the urban élite would for example gain priority over the development of a simple solar cooker for the rural population. Instances could be multiplied.

Governments need to develop scales of priorities based on human rights, of which the most relevant for our purpose are the rights to food, health, shelter, and a decent standard of living. These must take priority over the sophisticated technologies demanded by the élite. The competing claims to technology of these two groups are often lost sight of when technology is received or accepted by developing countries. It is important that the human rights aspect be given a higher profile, so that decision-makers in receiving technology will bear in mind this dichotomy of demands within the same system.

The Computer as the Generator of New Products

To all this must be added the fact that the actual processing of information, as well as the dissemination of information – not to speak of the storing of information – requires intensive investment in computerization. These again are largely within the control of one sector of the world's economy and it is that sector which thus acquires increasing control over the product and the information embodied within it.

The computer is no longer merely an assembler of information. It is today extensively used to generate new products and tailor them for particular purposes and specific markets. This resource, extremely expensive and beyond the reach of most developing societies, gives the developed world an irreversible lead in the ability to design a product for the markets of the developing world, so that there can in fact be little real competition.

The future will see the further spread of new computer techniques, including that of robot technology, which can have severe repercussions. Such technology could thus in effect deprive a developing economy of the use of its labour, which is one of its most important economic resources. Important policy decisions are called for in this area, in order to increase the ability of developing countries to make an input at the designing stage into the new technology.

Education for Participation

A steady policy aim of generating increasing technological skills in the local population is an important part of any policy aimed at turning technology to the service of human rights. Scientists and technologists from the country itself are likely to be far more aware of pressing local needs and demands than those from abroad. With more such skills available locally there will also be a greater trend towards discrimination in the acceptance of technology and towards taking apart a complex technological package rather than accepting it in its totality without question.

An attitudinal change is required here in the entire educational structure – a change that will accept technology not as a foreign graft but as an integral part of the local culture. Marrying traditional local technology to the new will have important psychological overtones for this process. The dichotomy between the traditional way and the technological way, which leads to the us/them syndrome regarding technology in developing countries, needs to be broken down. We are one world, and technology is the common inheritance of this one world, without which no part of it can survive. Just as Japanese culture integrated Western technology into its way of life to the extent that there is now nothing "foreign" about it, so also all developing countries must integrate technology.

The educational process must aim not merely at producing scientists and technologists who can actively participate. It must aim also at the general population, with the intention of making them more receptive to the technologies that will assist them.

Reaching the Grass Roots

Any process of educating the community must take note of the current deficiencies in the technological communications network.

Even within the UN system, despite its enormous output of publications, there is an insufficient targeting of "grass-roots" audiences. For example, a recent survey found that of WHO's list of approximately 1,300 titles, only 11 titles were suitable for primary health workers.[17] Although there were 1,800 publications (including periodicals) issued within the UN system in 1981, the average print run was only 2,000 copies per publication, despite the fact that the seemingly large number of 3.6 million volumes and journals was thus produced.[18] Of this limited output around 80 per cent went to the industrialized countries. For WHO the figure was 80 per cent, for ILO 78 per cent and for the UN itself 91 per cent.

Such figures underline the necessity for new concepts and methods relating to publishing for the grass roots. It is to be remembered that publication is not the only route by which the grass roots can be reached. Other alternatives that have been satisfactorily employed are:

1. *Demonstrations.* This is especially useful with small-scale technology. Nothing succeeds like demonstration of a good example.
2. *Prototypes.* It is useful to start a prototype of a new scheme at the grass roots and monitor its progress. Material can then be put out showing how the prototype can be replicated, drawing on the experiences in that very community.
3. *Promotion of the idea and of adaptations.* This is a process that must be actively undertaken with literature at the grass-roots level.

Academic publications, while they have their own merits, are no substitute for revitalized attempts to take technology to the grass roots.

Appropriate Technology

This is a topic that has generated a vast literature and there is no space here to give it the attention it deserves. Schumacher's *Small is Beautiful* is perhaps still the outstanding work in this genre, which has now grown to thousands of titles and a flood of articles, reports, and symposia. Since appropriate technology aims at adaptation of technology to the needs of the recipient country, the local input into the actual technology is considerable. This has the additional advantage of preventing a sense of alienation. Mahatma Gandhi's approach was not against machines but against man becoming mechanized. With appropriate technology, which is adapted to the service of the receiving community, there is little chance of man becoming the servant of the technology.

Appropriate technology does not necessarily mean that it should be entirely the adaptation of the receiving country. Some developed countries have active programmes for contribution towards the evolution of appropriate technology, as for example the German Appropriate Technology Exchange (GATE) under which various German agencies joined in developing a service package for developing countries. It concerns itself with the planning and implementation of technical co-operation projects dealing with the development, adaptation, transfer, and propagation of appropriate technologies. Developed countries can assist considerably in this way.

Schumacher's vision of a new method of production, rooted in what he calls Buddhist economics, which should be simple, non-violent, kind to the environment and not aimed at the stimulation and satisfaction of superfluous material needs, has many ideas within itself which can assist in turning technology to the furtherance of human rights.

In the midst of all this discussion we must not lose sight of the fact that education is a major lever for development. There must therefore be a plan to spread among each community an understanding of the way in which technology can be of assistance to it in its day-to-day problems. It is true there are already rather loose organizations aimed at spreading this sort of knowledge among the populations of third-world countries. There needs to be some co-ordinated effort, the guidelines for which may perhaps be formulated by the Human Rights Commission, pointing to the way in which each government could maximize the communication system in relation to technology that is appropriate for its local communities.

ENJOYMENT

The purpose of all the decisional and participatory processes thus far considered is of course the enjoyment of the technology by the maximum number of citizens of the country concerned. While the technology benefits society generally it must also be capable of individual enjoyment. It must not in other words be a technology which only the privileged can enjoy.

In deciding which of various sectors of a country's population should have priority in a given area of technology, perhaps a computation on the lines of a Benthamite calculus of utility could be employed. The technology must be capable of enjoyment by the largest possible sector of the population.

This is perhaps the most important of the trilogy of requisites within the General Assembly's definition, and we need to examine the serious obstructions impeding the movement of science and technology from the point of its creation to the point of its delivery to the individual member of society who is to enjoy it.

In this section I propose to examine some of these major obstacles. They include current attitudes of the scientific establishment, the tendency of the information technology to concentrate and entrench the technological dominance of the Western world, the armaments trade, the politics of food, and the inadequacies of the law in keeping pace with technology. The list is by no means exhaustive and I have selected only a few major areas. I shall consider each of these in turn.

It will be noted that some of them have relevance also to the earlier topics of participation and contribution.

Attitudes of the Scientific Establishment

Scientific Priorities

If we are to turn technology to affirmative service in the cause of human rights, we cannot even hope to achieve this result unless we can enlist the co-operation of the scientific establishment. Indeed, if we can, half the battle is won.

The Brandt Report, which stated that, while more than 50 per cent of the world's scientific manpower was devoted to the manufacture of weapons, less than 1 per cent was devoted to researching the needs of the developing world, makes imperative the diffusion of such knowledge. One begins to wonder whether there is not a responsibility lying on the appropriate international bodies to suggest some scale of priorities in relation to technological research. It is true there can be no enforcing mechanism in relation to priorities, but guidelines in this area would be of great value.

A disproportionate amount of resources is sometimes devoted to an area of research which benefits extremely few and may not contribute at all to a vast segment of the human race. Elaborate and expensive procedures for organ transplants or *in vitro* ferilization may prolong individual human life or result in the possibility of the creation of life in laboratory conditions. Such technology benefits the few and involves immense resources, thus leading us to the vital question of priorities in a world of shortages.

We need to work towards securing an acknowledgement from prestigious scientific organizations of the need to step up the amount of scientific endeavour spent on researching the technical needs of the developing world.

The universal eradication of malaria could be achieved at a cost far less than the rate of expenditure for one hour on the global armaments race. Common sense would dictate which one of these is the priority, but we tend to continue pouring money into the technological processes dictated by the needs of corporate, industrial, and military might.

Social Responsibilities of Scientists

Unfortunately there has not been a sufficiently sustained effort to bring home to scientists the human rights implications of their work. Scientists as individuals are as well intentioned as any other members of society, but their work has immense potential for the destruction of human rights.[19] All too often, however, the scientist, engrossed in his own particular research, does not see the social consequence of his work.

It is vital that a heightened consciousness of these social implications be fostered in scientists. This may be done through social responsibility courses introduced into all tertiary science curricula, where today they are singularly lacking, or by the introduction of ethical codes which different disciplines of scientists, e.g. engineers, would adopt. It could also be achieved through a wider diffusion of relevant information among scientists of the social consequences of their work, which international organizations such as UNEP (the United Nations Environmental Project), Unesco, or UNDP could stimulate. Indeed, there is room here for an international journal on the Current Social Consequences of Science, which one such organization, or even UNU, could undertake.

Consideration of this matter leads in turn to consideration of the responsibilities of the scientific community, who, after all, are in the vanguard of this process of diverting earth resources and human resources into wasteful expenditure. Scientists and technologists tend to give their expertise to the highest bidder. There can be no quarrel with this, but when the overall result is one which produces a degree of iniquity as great as that embodied in the 50 versus 1 per cent differential referred to in the Brandt Report, one begins to wonder whether the scientific community ought not to be devoting some effort to evolving a code of social responsibilities.

This is a much neglected area: except for the very rudimentary code of medical ethics centring around the Hippocratic oath, and some extremely rudimentary codes of the engineering profession or in the computer field, the vast profession of science is completely unregulated by ethical codes such as exist among lawyers and accountants. Consequently many of the world's most talented scientists devote their expert talent and skill to perfecting weapons of destruction, with no apparent qualms resulting from the perception that a fraction of their talents directed into different channels could save tens of thousands from starvation or malnutrition. In short, the scientific conscience needs to be stimulated.

The attention of future scientists should be particularly drawn to the imbalance in scientific effort already mentioned. They need to be shown how this may result in tensions so severe as to create dangers to world peace in the future. With the virtual monopoly of expensive scientific technology currently enjoyed by one sector of the world economy, the imbalance in the relative economic positions of the different sectors of the world is heightened to acute proportions, and may result by the turn of the century in the bitterest conflicts, conflicts that will unsettle the rich and poor world alike.

We should perhaps be considering also the inauguration of centres where international co-operation between scientists can be generated by drawing together scientists from different sections of the world for a discussion of these socially oriented problems.

The Perversion of Science and Technology

Reference should here be made to the well-known Poona Declaration on Scientific Responsibility, which describes itself as an indictment of the perversion of science and technology, and spells out in quite explicit terms some of the problems which this project would encounter. Among the items mentioned in the Poona Declaration, which was adopted by the participants at the fourteenth meeting of the World Order Models Project held in Poona, India, in July 1978, were those related to biological farming in the third world by pharmaceutical transnationals, the banishment of a growing number of first-world poor from productive activity through increasingly capital-intensive technology, and the employment of 50 per cent of all research scientists in the world in military research and development. The resolution urged serious reflection and a vigorous debate on the present predicament and on the need for an active search for alternative perspectives on science and technology, relating both to the pursuit of truth and the process of human liberation.

The Promotion of Humanistic Science

The development of a more humanistic approach to science and technology could be stimulated through the establishment of centres for the study of the human aspects of science and technology or centres for humanistic science. This is a needed corrective to the compartmentalization of science, which in this generation has seen scientific expertise fragmenting into ever smaller specialties and subspecialties. The deeper one goes into a narrow area of science, the narrower becomes one's outlook, and the less one communicates with other disciplines or even with the subdisciplines within one's scientific field. The perspectives that ought constantly to be before the scientist hence tend to be shut out. The remedy for this is to see science in an overall or holistic context. Apart from the injection of social perspectives into existing science curricula, there is a need for the establishment of centres devoted to this holistic approach.

Such a holistic approach would necessarily involve interdisciplinary studies and would bring together sociology, economics, the humanities, philosophy, jurisprudence, and many other areas of human knowledge that can make a direct contribution to fostering in scientists and in the scientific product a greater regard for humanity and the environment, which in the last resort scientists and science are intended to serve. Many of the problems concerning the impact of science on society and in particular on third-world societies are the result of failure to have this broader dimension prominently placed before the scientific enterprise.

A prototype for such an enterprise could well be the Mitsubishi Kasei Institute of Humanistic Science, a large research centre based upon this holistic and humanistic approach to science.

The informational aspects that go into the creation of new products need this orientation, and it is most important that these perspectives be fed in prior to the generation of the product — hence the importance of the humanistic approach at an early stage, before the technology is actually created. The impact of such a new attitude will be felt also in the field of scientific method, where results are currently determined by experiments tightly controlled within their own narrow framework, without regard to the broader social dimension. Results that fit in with the postulates of that narrow frame of reference are not results that will necessarily sit easily within the postulates of a broader social frame. The framework of scientific research must, in other words, be considerably broadened to take on a humanistic dimension.

Information Technology

A second great roadblock on the way to the bending of science and technology to greater human service is the way in which information technology is helping in the polarization of developed and developing worlds. Although one would expect that the explosive growth of information technologies in recent years would turn out to the benefit of the average citizen in the developing world by making a knowledge of science and technology more freely available in quarters that it would not otherwise have reached, the information revolution has thrown up yet another obstacle to the delivery of appropriate technology to the third world.

It is to be remembered also that whereas earlier technology, dating back to steam and electricity, greatly magnified physical power, modern information technology greatly enhances man's intellectual power,[20] and therefore greatly magnifies its possessor's powers of domination.

The Information Component of New Technology

It will be observed that the generation of information that is the prelude to the creation of a new scientific product often takes place through the availability of resources in the developed world, resources that few developing countries possess. These resources consist of trained researchers as well as an expensive apparatus for the collection of data. Although the collection of this raw data has to be undertaken in distant places, it is often processed in the rich world at centres far away from the sites where the information is collected. This is the first stage in the surrender of control by the developing world of knowledge concerning itself that could assist in the creation of a future product for use in its own territory.

Much of this information bears upon the increasingly recognized principle that international agencies are striving to protect under the head of transborder data flow. If knowledge is power, knowledge of local conditions, needs, strengths, and shortcomings is power over the society in question. Such data, once it travels beyond the boundaries of the country where it is collected, ceases to be under its control. The first stage in the fashioning or introduction of a new technology is thus placed beyond the control of the recipient country. It is vital that this question be addressed, bearing in mind also the proposition that knowledge is free and that research should be unfettered. There is here an important dilemma to which we should address ourselves.

The scarcity in the developing world of the two resources earlier mentioned, namely the human resource of the skilled researcher and the information-gathering apparatus, is one of the root problems creating this dilemma. It is possible that it could be addressed by devising an appropriate combination of resources from the two countries mentioned – the giver and the recipient of the technology – rather than there being a total surrender of control at this early stage to the giver. In other words, is it not possible to introduce a mandatory requirement of participation in the project, with property in the resulting information being shared between the two countries?

Even before the manufacture of the product resulting from the information thus generated, it thus already embodies property (in the form of raw or processed information) that belongs to the developed world, although its eventual target is the developing world. When the technological input takes place, the totality of the product becomes the property of the developed world and under current concepts of property and free trade it can do with it what it pleases and sell it on terms it alone determines. However much it might be needed in the developing country, barriers then arise between the product and the place in which it is to be used.

These barriers take the form of protection of intellectual property, protection of financial investments in the generation of the new product, and political barriers which need to be penetrated before the product can really serve the people for whom it is intended. These political barriers take the important form of levers of social, political, and economic control which are manipulated almost exclusively from the developed world.

Various aspects of human rights become relevant to a consideration of the resulting problems.

Clashes of Human Rights Principles

On the one hand there is a recognized right of the creator to a product, whether material or intellectual, and to the profits flowing therefrom. On the other hand there is a general principle that science and technology are primarily intended for the use of humanity and that information is free. Other clashes of human rights principles occur when we consider that the right of property is heavily protected under the Western jurisprudential tradition, while in the traditions of the developing world considerations of social interest and of humanity occupy a higher position in the hierarchy of values. The limitations that must be placed in the social interest upon the absoluteness of ownership principles tend therefore to be blunted where the dominance of Western technology gives dominance to the Western position.

These conflicts are not easy to resolve, but there are some trends in human rights jurisprudence that can be called on. One of these is the growing emphasis upon the concept of universalization of human rights norms. Concepts of absolute rights of property that belong to an exclusively Western jurisprudential tradition find it increasingly difficult to maintain their regime unimpaired when confronted universally by rival human rights traditions that do not accord this sacrosanct importance to the concept of property.

Another is the increasing importance of social, economic, and cultural rights, as compared with the civil and political rights which formed the kernel of the Western human rights tradition. All of these result in the progressive attrition of property rights in the interests of social welfare.

The matter can be illustrated by a contemporary example. It has been advertised recently that a medical product which prolongs the life of an AIDS sufferer will shortly be on the market at a cost of approximately $20,000 a year for each user. We do not know how authentic this claim is but it offers a textbook illustration. Such treatment may well be beyond the financial reach of a person who desperately needs this drug. We can think in a similar context of a hypothetical cancer cure which is marketed at $1,000 a pill.

Such examples highlight the clash between two human rights – the property right of a creator in his intellectual creation and the right to life and health of the user. Each is valued in its own context but when posed in opposition to each other there is an inevitable subordination of the right which has less muscle behind it.

Such examples, though far-fetched, draw attention to the dilemma that constantly faces the international community in relation to expensive technologies which the developing world needs desperately but which can only be obtained at a price beyond their reach. While we must have due regard for the expense, risk, and expenditure of human resources involved in the generation of the product, it is not unreasonable to argue that the product once achieved belongs to all humanity, subject to a reasonable compensation to the generator of the product for the effort and expense involved.

Scales of Priorities of Human Rights

The clash of principles outlined above points to the need for international guidelines to be evolved and brought into play so as to achieve a workable reconciliation between conflicting principles.

For example, it may be possible to define a set of guidelines under which products that are vital to life, once generated, become universal property, subject to a right of compensation at a level deemed appropriate by an international authority set up for this purpose. Perhaps the analogy of non-derogable rights which we can draw from the learning regarding derogation from human rights principles[21] could afford some guidelines as to human rights which are considered of such a compelling nature that they override others in the event of a clash. The funds for such compensation payments may not be available for those who need the product and it may be necessary to build up a buffer fund which can in such instances make the producer a fair reward for his initiative, effort, and expenditure.

This is not to say that every product thus generated will become the subject of such free transfer. The principles outlined above will apply in exceptional instances, such as are defined by the appropriate authority, and the justification for this would be that the rights to health and life stand above all other rights.

The writer is conscious of course that such a principle can operate as an inhibitor of research. It may well be that research organizations, particularly drug companies, would not invest the current level of effort in potential products, if they knew there was a possibility of the appropriation of the resulting knowledge into the universal domain. On the other hand, it is only in exceptional cases that the principles outlined would be brought into operation, and in any event there would be a reasonable costing of the effort and expense involved, so that loss would be averted and there would be a reasonable profit for the outlay.

Another consideration to be borne in mind is that the profits currently made through such technologies are well in advance of what may be described as reasonable remuneration. Pharmaceutical companies are well known for the very substantial profits they make, far exceeding what is considered, even in the competitive financial world, a reasonable remuneration for the outlay of resources. It may well be that similar principles apply in regard to new technologies that generate agricultural products.

The collection and processing of information, which, as we observe in this section, are an essential prelude to the creation of a product, thus involve not merely the collection and classification of facts but also an analysis and evaluation of any clash of principles involved. Seeing that the informational input prior to the creation of the product will be of increasing importance as the basis of scientific and technological development for the foreseeable future, the problem we are confronted with will, if at all, become more acute. We need to address it now.

The Role of the State

It is clear that the state has an important role to play in achieving the maximum enjoyment of technology by its citizens. One of the most important problems faced by developing countries is the nature and extent of this state role.

It would be of assistance to examine in this connection a pioneering Sri Lankan project, the Million Houses Programme (MHP). This programme offers both an interesting philosophical base and a practical example in fulfilling the basic human right to shelter of a large segment of the population. At the thirty-fifth General Assembly of the UN in 1980, Sri Lanka proposed the concept of a selected year to focus on shelter, which resulted in the UN's International Year of Shelter for the Homeless, 1987. The rationale behind the proposal, as stated by Prime Minister Premadasa to the UN (12 October 1987) was that great industrial and agricultural visions tended to by-pass the centre of human development – the home.

The programme, which has already achieved the construction of half a million houses, rests upon a support-based paradigm (S-BP) rather than the traditional provider-based paradigm (P-BP). This means that it is poor families and poor communities that both *decide* and *do*, the role of the government being to support and complement their initiatives. Under this philosophy all technology issues are individually and locally determined by the micro-context rather than determined from above in conformity with a macro-scheme. In making known the technologies available, the government has an important role to play, as it has in providing credit, basic services, training, and technological assistance. Housing options and loan packages (HOLPs) facilitate choice of technology and materials. Yet the underlying philosophy is that the state participates in the people's process rather than the people in the process of the state. State support is maximal; its intervention is minimal. Technology choices are for the people and individuals of the area. Their range of information is considerably magnified by the services provided by the state.

We refer later in this chapter to the importance of self-help. It may well be that there are principles of global significance in this regard in the Sri Lankan experience.

One-way Information Flow

Another area of marked imbalance in the current world order is the imbalance between the resources of the developing and the developed worlds in relation to news flow. This may not at first sight be seen as a means of perpetuating technology differentials, but it *is*, in fact, especially when one considers that the entire global communications network, as currently organized, presents a largely one-sided information flow.

Raw information may come in from the developing to the developed world but the processed information travels out in a one-way stream of traffic from the developed to the developing world, while nothing that can match it flows in the reverse direction. In order to improve the human rights of the vast sectors of the world population thus disadvantaged, it is necessary that telecommunication facilities and wire services be available to the developing countries, with a sufficient global spread to relay from them processed news information to all sectors of the globe. The lack of availability of such a system means that the human rights of this vast group are impaired, for the entire world receives its news and forms its views on the basis of the one-way flow of information generated in the developed world. This is in regard not merely to information which we call "news," but to all species of information, on the basis of which needs are perceived, policies are made and products are designed. In the result the problems of the developing world are not seen by the world public in the real way in which they should be seen, that is in their overall context, and the assessment in the developed world of a third-world problem or need tends often to be distorted, not for lack of goodwill but for lack of information.

This means that in the resulting political process or in the resulting economic transactions, the legitimate human rights of such developing-world populations tend to be negated or whittled down to an extent not permissible under current universal perceptions of human rights. Unesco at one stage considered in detail this imbalance of news.

We need also to consider the imbalance of technologically related information. The impact of technology on development and developing societies cannot be considered adequately without due reference to this aspect, which forms an important segment of the area of research that this project should cover.

The Armaments Trade

Another dimension of the problem under review is the interconnection between disarmament and development. Although the intertwining of disarmament and global development is well acknowledged in theory, very little has been done in practice to achieve the desired result of diverting some portion of world resources currently wasted on armaments into the field of development. This is an area involving superpower and regional politics, but also one to which the Human Rights Commission may have some worthwhile contribution to make. Indeed, the Commission would perhaps fail in its duty to consider the overall picture if it did not give some attention to this vital aspect of the development dialogue.

There have been insufficient analytical studies of the ways in which the arms trade conflicts with nearly every canon of the current universally accepted body of human rights. This can be written on *in extenso*, but it is not within the scope of the present study.

However, an important area of the arms trade which needs continuing in-depth research is the question of diverting the productivity of the armaments industry into products of a peaceful nature. Conversion of swords into ploughshares is perhaps the most important and urgent of the studies that need to be undertaken. Considerations of space prevent a more detailed study of this area, which is rapidly becoming a major area of research.

The Politics of Food

A major obstacle to the enjoyment of modern technology by a vast segment of the world's population in a vital arena is the fact that food has become the plaything of a political-economic-technological complex that works its equations of profit ratios far from the starvation and malnutrition produced by the maldistribution of food. These human problems are seen almost as an academic exercise, far from the scene of the problems themselves. All the resources of computer technology and satellite scanning are used to manipulate the futures market in grain, and the ledgers recording these transactions show only the number of dollars gained rather than the number of human lives lost.

Thus, although through the aid of technology world cereal production has steadily outstripped world population growth, and although the world produces more food per head of population than ever before in human history (in 1985 it produced nearly 500 kilograms per head of cereals and root crops),[22] more than 730 million people did not amidst this plenty eat enough to lead fully productive lives.[23]

This increase in food production is of course the result of improved technology, some of it of the most sophisticated kind. Satellite imagery, microelectronics, computer sciences, biotechnology, tissue culture techniques, and other frontier technologies help produce this result. These technologies are, however, under the control of the affluent world. Its grain traders can put together a forecast of the world's crops by using satellite infra-red scanning procedures for forecasting a bad wheat crop – e.g. in the Soviet Union – thus assisting in the manipulation of futures markets. Its politicians can use the supply or withholding of grain as political weapons and can decide when its farm production of surplus grain is to be curtailed or stocks are to be destroyed to bolster up prices. Its industrial-technological complex can offer or withhold a new grain production technology to a third-world government or ruling élite.

In the decisions that are taken the interests served are of course not those of the bulk of the world's population but the financial and political interests of the owners of the technology. Else we would not see, for example, vast stocks of food being burned or cattle being slaughtered. Nor would we see massive cutbacks in agricultural production made mandatory by law in great grain-producing countries such as the US or Australia.

The fact that food production and technology go hand in hand in the 1980s points to the need for food technology to be harnessed in the service of that greatest of human rights – the right to life. Food adequate to the maintenance of a life of dignity is a human right, and technology needs to be harnessed in the cause of food security.

All of the technologies we have mentioned and many more must be tapped to meet the needs of the developing world, but the obstacles caused by commercial attitudes of profit-making, scientific and technological lack of concern, and feelings of futility and resignation on the part of third-world leaders and scientists must be overcome.

The solution lies clearly not in the handing out of food bonanzas to the developing world on a regular basis, but in their greater participation in such technology. To this end we need revised attitudes of the ruling élites, greater scientific and social awareness, and international norms which regard food as a public resource rather than a purely private right.

The following areas are suggested as needing increased attention:

1. The development of norms of national and international law in the field of intellectual property, seeking to build up within the universal domain a pool of scientific and technological knowledge. By way of example, private companies seeking property rights to improved seed varieties tend often to fail to recognize the rights of the countries from which the plant matter was obtained. As the World Commission on Environment and Development pointed out (p. 139), such practices could discourage countries rich in genetic resources from making these internationally available, thus reducing the options for seed development in all countries. Norms of international cooperation can also be built up in these areas, just as such norms have been built up in the field of environmental law.

2. International organizations such as FAO already hold an unrivalled bank of agricultural statistics and inventories of the world's physical resources. Such organizations can draw in the most sophisticated modern technology to help the third world. An outstanding example is FAO's satellite-based alert system, backed by scientific establishments in the Federal Republic of Germany, used to give early warning of drought, crop failures, and insect plagues in Africa, thus enabling preparation for major food shortages before they occur.[24] Databases are being built up covering agro-climatological statistics, day-to-day vegetation indexes, and an analysis of potential locust habitats in Africa, the Middle East and South-West Asia.

3. Within the developing countries there must be a greater readiness to set up and finance research institutions that can help in channelling the new technologies towards the needs of the country in question. Many third-world countries, knowing their weakness in research, tend to take this situation for

granted. This further increases the technological gap with the developed world instead of providing the technological bridge we need.

4. More specific international guidelines and national laws are required for regulating food technology that is introduced into a developing country.

5. International lending institutions such as the World Bank and the Asian Development Bank can lay down firmer guidelines requiring the development of local technological skills in relation to agriculture and fisheries.

6. The world community, through its resources and encouragement, must seek to divert towards food technology an increasing segment of the world's scientific skill, of which, according to the Brandt Report, less than 1 per cent is devoted to the needs of the developing world. The vast bulk of scientific research on agriculture still concentrates on the needs of the developed world, whose resource-rich conditions and ample water supplies contrast with areas such as sub-Saharan Africa and the remoter areas of Asia and Latin America with unreliable rainfall and poorer soils. These call for new and specialized technologies suited to each particular area, which would represent a combination of the modern and the traditional.

7. The food problem is linked with the land problem and research is required to free the maximum amount of land for food production. Land law reforms would often have to be effected to meet this need. The necessary studies would include the disciplines of law and economics, having regard in particular to the fact that much of the most serviceable food-producing land in the developing world grows produce for the developed world rather than food for its own population.

8. Freeing more land for food production involves environmental concerns as well, such as rehabilitation of mined-out land. Many new technologies are available for this, and they need to be harnessed, while, at the same time, the principles of environmental law need to be developed.

9. Human resources need development through educational processes, leading to a greater receptivity to modern technology – e.g. hydroponic cultivation, biomass generation of energy, use of wind power for irrigation, solar energy for such simple purposes as cooking, compost production in place of imported fertilizers, crystal distribution techniques for enhancing the water-retention capabilities of soil. Many of these technologies, often thought of as remote and sophisticated, are well within the reach of the average third-world peasant. Many new products such as fertilizers and pesticides are misused for lack of necessary technical knowledge. In short, a wider spread of technological knowledge both at the level of the research institute and the average farmer is required if the politics of food as an exploitative weapon is to be blunted and the technology of food production is to be brought into the affirmative service of promoting human rights.

10. A number of points concerning ethical aspects of the right to development were stressed in the 1978 UN study on development as a human right.[25] Since science and technology are a major component of the concept of de-

velopment, every one of these propositions would apply to science and technology and the sharing of such knowledge, which is universal in its nature. There is thus a firm ethical base, soundly grounded in international law and its evolving norms, for the proposition that the shrinking world in which we live needs to be protected, for our common benefit, by treating science and technology not as a private preserve but as a universal inheritance.

Inadequacies of the Law

The dynamic growth of science and technology has outstripped the ability of the law to control it. Both legal procedures and legal concepts have proved unequal to the challenge.

Any study of legal inadequacies would have to concentrate on four important aspects: structures, concepts, procedures, and personnel. Quite often all of these in colonial and post-colonial contexts still bear the stamp of their colonial origins, and are thus suited basically to the societies of the West.

A formalized legal structure, with formal and expensive procedures and legal concepts tailored to the needs of the individualist, property-owning Western society, are hardly adequate for guarding against technology's potential for damaging human rights or for assisting in the transference of appropriate technology to the third world. This is an important roadblock on the way to turning technology to the affirmative service of human rights in those societies.[26]

Even in Western societies courts are recognized as not being suitable agents for the monitoring of technology or for determining in advance whether a given technology is suitable. Yet such questions tend often to come before the courts. Other agencies are required for this, as, in general, a court looks back at a past event rather than forward to the future. Value judgments, resource allocations, and community priorities are not appropriate matters for courts. Moreover, judges are not trained in science and technology, and therefore lack the expertise with which to judge the increasing proportion of technology-related disputes which come before them.

Against this background, the inadequacy of formal court structures and procedures to meet third-world conditions becomes even more apparent. Adversarial litigation does not provide for representation of the public interest but only for that of the two contending parties. It produces a finding which is effective only between the two parties and is not necessarily related to social needs. It aims at finding which of two parties should win by the rules of the game rather than at ascertaining the truth. Its procedures are expensive and dilatory, whereas technology often requires quick perception and action. Evidentiary rules are often archaic and exclude for formal reasons many considerations which could be of real value in fact-finding. The thalidomide case provided an excellent example of the inadequacy of old procedures to deal with a great social issue resulting from technology. Thus, a decision of the British House of Lords had to be discarded recently by the European Court of Human Rights in order to enable publication in the public interest of material which the British courts had shut out on grounds of legal procedure and privilege.

Old concepts of the law are likewise proving inadequate. Concepts of privacy and physical trespass were worked out before the days of surveillance devices, "bugging," and computers. Laissez-faire principles in scientific research are still accepted without limitation, despite the public interest in science and technological matters. Concepts of unrestricted private property are still unthinkingly applied by courts and judges.

Many third-world societies still believe that their courts and legal systems can carry burdens in the scientific and technological age which they were never designed to bear. In this belief they refrain from constructing the new structures and instrumentalities which are needed to monitor technology and make it harmonize with the needs of people. The existence of this obstacle must be more widely realized if science and technology are to serve human rights more affirmatively in third-world societies.

It will be necessary for a special study to be made identifying legal obstacles that stand in the way of the realization of these objectives. Work has already been done in relation to development by such centres as the International Centre for Law and Development to identify legal obstacles to the realization of alternative development strategies. This work has studied the ways in which those who desire to preserve the status quo on the developmental and technological fronts use law as a means of obstructing change.

The problem is both a national and an international one, and the study undertaken will thus have to be in the fora of domestic law as well as international law. There will also need to be a study of the ways in which co-operation can be achieved among domestic legal systems in eliminating these obstacles. More specifically, studies will be needed of ways of building into legal structures in developing countries a greater control over production and the processing of information. Other areas where greater legal control will be required will be in relation to control of trading relations and control of prices.

In the last section of this chapter we consider the reorientation of lawyers' attitudes that will be necessary for this purpose.

SOME DESIDERATA

The Human Factor

In a programme on human rights and development we cannot afford to lose sight also of the central fact that in all development processes one of the most important aspects is the human factor. In fact this has sometimes been described as the missing factor in development because it tends so easily to be overlooked. This human resource consists of the basic skills of the workforce, the administrative skills, the social structure of the rural sector, and the expertise already contained within the scientific and technological communities, however small, in these countries. It is important that all these resources should be marshalled as part of the general effort towards making these societies receptive to the better impacts of the new technologies.

It is to be remembered also that, although in Western countries with a tradition of three or four centuries of industrial development there has been built up over the centuries an appropriate infrastructure for the spread of technology, in the developing world this long process is now being compressed within a very short time-frame. If the beneficial effects of the new technologies do not reach those who can best benefit from them, this is a principal cause. Much of the effort that we shall be putting into the use of new technologies will be dissipated in the absence of due attention to the human-resources aspect of development.

Traditional technology, rich especially in relation to agricultural activity, embodies the wisdom of generations. The tendency to cast it aside without due examination under the impact of modern technologies, such as high-level fertilizers and Western machinery, should be resisted. This is a rich human resource which needs to be considered and sufficiently used.

As the Brandt Commission observed after a wide-ranging survey of the concept of development:

This Commission did not try to redefine development, but we agreed (among other things) that the focus has to be not on machines or institutions but on people. A refusal to accept alien models unquestioningly is in fact a second phase of decolonization. We must not surrender to the idea that the whole world should copy the models of highly industrialized countries.[27]

Raising the Political Priority of Development

Another aspect to which the Human Rights Commission should give its attention is the means of raising the political priority of development, both in the developed world and in the developing world. In the developed world there is a strong feeling that the assistance given for the process of development finds its way into the wrong hands and that much of it is frittered away. Consequently there is the feeling that money spent on such forms of assistance is not money that truly filters into the process of development. In the developing countries themselves there is a sentiment developing that assistance given in this way is in fact a means of increasing economic and political control over the recipient country. An effort needs to be mounted to ensure that both these criticisms will be met by appropriate corrective action. This can assist in countering the growing resistance now discernible in the developed countries, both in government and in public attitudes, towards the giving of assistance which does not really reach its target beneficiaries. Moreover, if assistance in relation to technology comes without economic or political strings, its acceptability and impact in the developing world will be all the stronger. Studies of the extent to which development assistance is frittered away or made contingent to benefits to the donors will be useful pointers to the way in which these abuses can be minimized in the future. With a reduction of these abuses the political priority of development will rise.

At the same time the communication apparatus which is now growing up in the developing countries can also be used to spread, not only among political leaders but also among the people of those countries, the need for development aid to be received in an independent fashion, without political or economic strings attached to it.

Self-help

Self-help groups need also to be generated, not only in local areas but also in relation to particular industries. The communication technology available already in developing countries has not been used sufficiently for the purpose of generating the necessary concern and interest, which would result in viable self-help groups. There is a feeling that nothing one can do will halt the inevitable, and the resulting frustration needs to be countered. Demonstrations of the ways in which new technology can be used and new products generated need also to be taken to the rural areas, as well as to the workplace. A new work ethic and a new self-confidence must emerge.

Food as a Human Right

Work needs to be done in the human rights field to elevate the status of the right to food. The UNU has done promising work in this regard. Approaches to the problem of hunger have taken many forms, and a useful summary of the different approaches is contained in the chapter by Thomas J. Marchione in the UNU volume, *Food as a Human Right*. The five approaches set out here are the epidemological approach, the ecological approach, the econometric approach, the structural approach, and the advocacy approach. In each of these there is a technological element, and the Commission would perhaps like to consider the ways in which the technology relevant to each of these approaches is harnessed in the course of making available this basic human right.

Training of Personnel

The training of personnel for development purposes is an important facet which needs attention. This may seem too obvious to stress, but one criticism that can be made of many of these programmes is that they tend to be European- or American-centred. It takes time for a selected trainee to acclimatize himself or herself to the social and linguistic milieu in which he or she is to receive training. One area that needs to be explored is the way in which training courses can be planned within the region concerned. This will not only have the advantage of making the trainees feel more at home but will also keep them closer to the immediate problem and enable an exchange of experiences with those who are similarly placed. The tendency to depend on foreign expertise dies hard.

Law Reform

It is important that the services of lawyers in developing countries be enlisted to serve the needs of turning technology towards the furtherance of human rights. There is much that needs to be restructured in the fields of legal concepts, procedures, and structures. Moreover, the attitudes of lawyers themselves need to be reoriented, for their thinking is often cast in a predominantly Western mould by reason of the Western-oriented training they have received.

All this is vitally important to development, as one of the principal roadblocks on the way to development is presented by outmoded legal concepts, structures, and procedures.

In other words a new third-world jurisprudence must be developed. Hopeful steps in this direction have already been taken.[28] In the words of Professor Marasinghe,

It is arguable that one of the roles of law in development should be seen as the integrator of ecological, cultural, social, economic, institutional, and political dimensions of a given society, so that the diverse trends and aspirations in each of these fields could be synthesized in a way that the society as a whole could evolve into a cohesive social unit.[29]

In the words of a distinguished former President of the International Court of Justice, in his foreword to the same volume, the lawyer will have something of each of the following roles to play:

All these considerations lead us to envisage the possible role of law and the lawyer in this great task ahead of the nation. Is the lawyer, on whom the burden will likely devolve, to be just a utilitarian in search of the greatest happiness of the greatest number, an analytical jurist primarily concerned with concepts, a social engineer preoccupied with how law affects man's social interests, a functionalist anxious to ascertain the functioning of legal rules, a realist sold on the idea that psychology is of the essence of the judicial process, a Pericles engaged as a policy-maker wisely dispensing laws, or a plumber skilled only as a technocrat whose main interest is in "lawyers'" law?[30]

It is idle to speak of developing human rights unless we enlist the co-operation of the legal profession in each country, for lawyers can be a great help to the cause, just as they can also be great obstructionists.

CASE-STUDY PROJECTS

I would like to conclude this paper by referring to a very practical problem in which I happen to be involved and which clearly illustrates the importance of using modern technology for the purpose of promoting human rights. In the island of Nauru phosphate lands have been mined out by various occupying powers from the beginning of this century, leaving sizeable portions of the island with nothing more than bare coral pinnacles, very much resembling a moonscape. A large area of that little island is now unfit both for human occupation and for any other purpose, and the inhabitants of the island do not enjoy the basic human right of enjoyment of their natural environment. A Commission appointed by the Government of Nauru is examining ways in which modern technology can restore this barren land to human use.

This is eminently a situation where science and technology can be affirmatively used for the furtherance of human rights. Article 13 of the Charter of Economic Rights and Duties of States of December 1974 clearly states that every state has the right to benefit from the advances and developments in science and technology for the acceleration of its economic and social development. Article 9 of the same document points out that all states have the responsibility to co-operate in the economic, social, cultural, scientific, and technological fields for the promotion of economic and social progress throughout the world, especially that of the developing countries.

The situation in which the island of Nauru is now placed requires, then, that a co-operative effort be made by all nations to assist in bringing to its aid all the resources that modern technology can offer, not merely because it is an environmental problem involving the human rights of the people of Nauru, but also because it is an environmental problem which affects the people of the world generally – the sort of problem that could well exist or be duplicated in other societies.

It may be worthwhile to consider cases of this nature as specific case-study projects with a view to seeing in what way the immense benefits of technology can be harnessed towards the solution of a problem of great consequence to human rights. In addressing a problem such as this we may well find not only answers to the specific problem, but also answers in the form or methodology that can be used for the solution of others.

The above represents some of the ideas which perhaps may be pertinent to the tasks the Commission will have in hand. It would be too ambitious to undertake an examination of them all, but it may be that some at least of these ideas will be considered important enough to warrant adoption by the Human Rights Commission as part of its programme. It is believed by the author that attention to these problems will certainly assist in improving the total benefit that developing countries will receive from the vast movements in science and technology that are now taking place, with resulting impact upon all societies, especially those of the developing world.

NOTES

1. Hague Academy of International Law/UNU, *The Right to Development at the International Level* Sijthoff & Noordhoff, 1980), p. 94.
2. K. Sharma, "In Search of an Alternative Science," *Development and Co-operation*, no. 4 (1987): 26.
3. See, generally, S. Goonatilake, *Crippled Minds: An Exploration into Colonial Culture* (Vikas Publishing House, 1982), chap. 9: "Science, Culture, and Social Systems: Some Theoretical Impressions."
4. See the statement at the 14th meeting of the World Order Models Project in Poona, India, 2–10 July 1974, titled "The Perversion of Science and Technology: An Indictment," in R. Falk, S. Kim, and S. Mandlovitz, *Toward a Just World Order* (Westview Press, 1984), p. 359.
5. For a reference to such a class in Africa and Asia and its links with multinational corporations, see Louis Turner, *Multinational Corporations and the Third World* (Allen Lane, 1973), p. 67.
6. "Neither Prince Nor Merchant: Citizen," *Development Dialogue*, vol. 1 (1987): 170.
7. See, generally, C, Cooper "Science, Technology, and Production in the Underdeveloped Countries: An Introduction," *Journal of Development Studies*, vol. 9, no. 1 (1972); F. Stewart, *Technology and Underdevelopment*, 2nd ed. (Macmillan. 1982), p. 278.
8. See, generally, C.G. Weeramantry, *The Slumbering Sentinels* (Penguin Books, 1984).
9. Thomas Land, "India Tightens Chemical Hazard Control," *Development and Co-operation*, no. 6 (1987): 18.
10. *Another Development for SADCC*, report of a seminar of November 1985 in Lesotho (Dag Hammarskjöld Foundation, 1987), p. 164.
11. A.W. Goldsworthy, ed., *Technological Change: Impact of Information Technology* (Australian Government Publishing Service, 1981), p. 56.
12. Stewart (note 7 above), p. 3.
13. Turner (note 5 above), p. 70.
14. Turner (note 5 above), p. 70–71.
15. See generally *Intra-national Transfer of Technology* (Asian Productivity Organization, Tokyo, 1976), chap. 2.
16. See Turner (note 5 above), p. 67.
17. Christopher Zielenski, "Reading the Grassroots: Publishing Methodologies for Development Organizations," *Development Dialogue*, vol. 1 (1987): 196.
18. See *Publications Policy and Practice in the United Nations System* (Joint Inspection Unit, Geneva, 1984) (Doc. no. JIU/REP/84/5).
19. See Weeramantry (note 8 above).
20. See *Technological Change: Impact of Information Technology* (Australian Government Publishing Service, 1981) p. 50.
21. See the Paris Minimum Standards for Derogation of Human Rights Norms in a State of Emergency (September 1984).
22. *Our Common Future*, Report of the World Commission on Environment and Development (Oxford University Press, 1987), p. 118, based on data from *FAO Production Yearbook 1985* (Rome, 1986).
23. See World Bank, *Poverty and Hunger Issues and Options for Food Security in Developing Countries* (Washington, D.C., 1986).

24. *Development and Co-operation*, no. 2 (1988): 24.
25. E/CN/4/1334, paras. 39–54.
26. For a fuller discussion see Weeramantry (note 8 above), chap. 3: "The Inadequacies of the Law."
27. *North-South: A Programme for Survival*, the Report of the Brandt Commission (Pan Books, London/Sydney, 1980), p. 23.
28. See in particular Marasinghe and Coklin, eds., *Essays on Third World Perspectives in Jurisprudence* (*Malayan Law Journal*, 1984).
29. Marasinghe and Coklin (note 28 above), p. 4.
30. Marasinghe and Coklin (note 28 above), p. v.

BIBLIOGRAPHY

Dag Hammarskjöld Foundation. *Another Development for SADCC*. Report of a Seminar of November 1985 in Lesotho. Dag Hammarskjöld Foundation, 1987.

Goonatilake, S. *Crippled Minds: An Exploration into Colonial Culture*. Vikas Publishing House, 1982.

Hague Academy of International Law/UNU. *The Right to Development at the International Level*. Sijthoff & Noordhoff, 1980.

Stewart, F. *Technology and Underdevelopment*, 2nd ed. Macmillan, 1982.

Weeramantry, C.G. *The Slumbering Sentinels: Law and Human Rights in the Wake of Technology*. Penguin Books, 1984.

——. *Nuclear Weapons and Scientific Responsibility*. Longwood Academic, 1987.

World Commission on Environment and Development. *Our Common Future*. Oxford University Press, 1987.

9

Human Rights and Environmental Issues

VID VUKASOVIC

The relationship between the development of science and technology and human rights has already been on the agenda of various United Nations' bodies for two decades. Different aspects of the problématique have been studied and many reports produced. The purpose of this paper is not to trace the history of the United Nations' work on the question[1] or to dwell at length on the human rights problématique, a field in which a rather rich literature exists. Its aim is to shed more light on the problem, primarily from the point of view of international law and the environment, having in mind the main goal of the study undertaken by the United Nations University in accordance with the invitation of the UN Commission on Human Rights (Res 1986/9).

HUMAN ABILITY TO DAMAGE THE ENVIRONMENT

We humans are, as Virginia W. Rasmussen put it, technological creatures and "tinkering is our nature."[2] Tinkering with nature is, unfortunately, not always done intelligently enough to avoid its impairment. People are not aware, or not aware enough, of the consequences of ecologically unsound activities. Many such activities, and primarily those linked to the improper use of science and technology, are going on in the world, causing deforestation, desertification, pollution of air, water, and land, damage to many plant and animal species and threats to other renewable as well as non-renewable resources of our planet.

Homo sapiens "has acquired the power to transform his environment in countless ways and on an unprecedented scale. Both aspects of man's environment, the natural and the man-made, are essential to his well-being and to the enjoyment of basic human rights – even the right to life itself."[3] W.J.M. Mackenzie wrote that "man had been too successful as an animal; . . . by his ever accelerating growth in numbers and skills he threatened his environment and therefore (as the laws of population ecology require) his own future as a species."[4] Man's ability to change his environment for better or for worse is, by the way, not a new development at all. Already prehistoric man could, by using fire for instance, drastically change his environment "intentionally or by accident." "The traces of man-made fires lie over the whole prehuman world."[5] So it is not strange that modern archaeology is concerned, besides other things, with ecology.

THE ECOLOGICAL AWAKENING

The ecological awakening of the early 1970s, which culminated in the United Nations Conference on the Human Environment, held in Stockholm in 1972, was the result of scientific research which proved that the state of our environment was to a lesser or higher degree impaired (in some cases alarmingly). The impact of this discovery was felt first in the academic community, but later on by others as well. Gradually, practically all scientists, most politicians and other decision-makers, as well as the general public, accepted the idea that our environment was in danger and that something had to be done. General legal principles (at least some of them) were relatively easily developed and accepted, at least formally. It was quite a different task, however, to undertake practical steps toward protection and improvement of the environment, on both national and international levels. Economic, political, military, and other interests often collided with the environmental issue. Besides, even scientists themselves had different views on the very complex problématique of environmental protection on one side and on the human rights problématique and other cognate domains on the other.

INTERRELATIONSHIP BETWEEN ENVIRONMENTAL REGULATION AND HUMAN RIGHTS

Our intention is certainly not to review the whole of environmental law, which has during the last two decades evolved into one of the most dynamic and expanding branches of international[6] as well as national law, but to reconsider only some problems which are more interesting than many others from the standpoint of the human rights problématique. It should be stressed, however, that all regulation in this field impinges directly or indirectly on human rights. It may be action against desertification, attempts to lessen or stop acid rain, better control of food production, efforts to make human settlements more habitable, or any other activity in the vast field of the environment, but it always protects or improves some human right.

Although this development has been rapid it is still inadequate in many fields, in spite of the fact that basic principles have been developed. Serious efforts at regulation in a more concrete way (by international conventions, by institutional arrangement, and in other ways) must be undertaken, on a global, regional, subregional, or bilateral basis. That means more concrete legal regulation at all levels, as well as additional work on elaboration of basic principles, and in the first place of the principle of "liability and compensation for the victims of pollution and other environmental damage," as formulated in the Declaration of the United Nations Conference on the Human Environment (Principle 22). Besides that, it should be kept in mind that states are responsible not only towards other states and the international community as a whole, but also towards their own citizens, who have the right to a healthy environment.

The enjoyment of all human rights is closely linked to the environmental issue. Not only rights to life and health in the first place, but also other social, economic, cultural, as well as political and civil rights, can be fully enjoyed only in a sound environment. And certainly, to go to an extreme, they cannot be enjoyed at all if the environment becomes impaired beyond a certain critical level. The whole of mankind could in such a case perish together with all its civilization, including human rights. The worse the environment becomes, the more impaired are human rights, and vice versa. That is the reason why there is the need for sustainable development and that means, in the first place, ecologically sound development of economies, science and technology, and all other fields. This is a *sine qua non* for both protection of the environment and further promotion of human rights.

Besides the undeniable interdependence between the environmental issue and all human rights, a new human right – the right to an adequate environment – is emerging. This right, still not precisely formulated, appears in documents and in literature, in some cases as a collective and in other cases as an individual human right.

THE RIGHT TO AN ADEQUATE ENVIRONMENT

The right to an adequate environment or, as it is termed in some texts, a satisfactory environment, is one of the so-called third-generation or solidarity rights. It can be found in international documents of both a declaratory and formally binding nature, as well as in domestic legislative and other acts of a number of countries, including some constitutions.[7] The African Charter, for instance, proclaims that: "All peoples shall have the right to a general satisfactory environment favourable to their development."[8] In the Declaration of the United Nations Conference on the Human Environment (Stockholm, 1972) it appears, however, also as an individual right. The Declaration states that: "Man has the fundamental right to freedom, equality and adequate conditions of life in an environment of a quality that permits a life of dignity and well-being and he bears a solemn responsibility to protect and improve the environment for present and future generations."[9] It appears as an individual right also in the report of the World Commission on Environment and Development, which proposes, as one of the legal principles for environmental protection and sustainable development, that:

"All human beings have the fundamental right to an environment adequate for their health and well-being."[10] Finally, it should be noted that elements of this right can be found in the Universal Declaration of Human Rights, as well as in both Covenants,[11] although the environment as such is scarcely mentioned in the documents.

It should be added that individuals, as well as groups, not only have the right to an adequate environment, but also the duty to protect and improve the environment. They have this responsibility not only towards other individuals or the community in which they live but towards mankind as a whole and even "future generations."[12] It is a responsibility which certainly collides in many cases with the enjoyment of their other rights, be they common citizens or those who, as scientists, technicians, decision-makers or in any other way, are more closely linked to scientific or technological development, environmental protection, health protection, or other cognate fields.

DEFINITIONAL PROBLEMS

Many questions arise in connection with this right, when it is formulated as a collective right, and some of these questions are similar to the questions concerning other human rights of the so-called third generation. In the first place it is not easy to see "how individuals can assert it against states, and so how it can be satisfactorily classified as a human right."[13] Besides that, we still do not have a precise definition of the right, whether as a collective right ("the right of all peoples," or of "people," or of all mankind, or of future generations) or as an individual right (right of man or of all human beings). Paul Sieghart believes that in the case of all "third-generation" rights, including the right to an adequate environment, "some formulation will have to be devised whereby each of them can be clearly seen to vest in individuals, to be exercisable by individuals, and to impose precise correlative duties on states so that it can then be interpreted, applied and enforced accordingly."[14] It is certainly true that further work on the "third-generation" rights is needed if we wish to improve the whole system of international legal protection of human rights.

Developments in the field of the environment, which are faster than in many other fields, will without doubt contribute to the establishment of a balance, as well as more links, between the first and second "generations" of human rights, as well as clarify the relation between individual human rights and the "third-generation" or collective rights. It is necessary to treat the whole problématique of human rights as a system in which all components are interrelated and play specific and functional roles. We stress the statement in the UN Declaration on the Right to Development that "equal attention and urgent consideration should be given to the implementation, promotion, and protection of civil, political, economic, social, and cultural rights and that, accordingly, the promotion of, respect for and enjoyment of certain human rights and fundamental freedoms cannot justify the denial of other human rights and fundamental freedoms."[15] Further development of the right to adequate environment will certainly contribute to these efforts.

One of the most important measures in the field of human rights is, without doubt, further work on a more precise definition of specific human rights, including the right to an adequate environment. Each of them must be explicitly defined, especially in national jurisdictions. While some human rights are relatively easy to define, others are not. The specific right to privacy, to mention only one example, "had not been explicitly defined in many countries."[16] Without setting precise definitions, efficient regulation by legal and other means is difficult, if not impossible, to imagine. It must be added that on the international level, when it comes to their definition, the rights which have newly emerged (like the right to adequate environment or the right to development) and emerging human rights deserve special attention. The work on definitions should be done not only by legal experts but by natural scientists, technicians, and other experts on a multidisciplinary basis. In many cases scientific and technical expertise is needed and various rules, such as ecological standards, are necessary as an addition to the more classical legal definitions and regulations.

THE COLLISION OF RIGHTS

The right to an adequate environment, by its very nature, is one of the rights colliding with other rights, in a certain sense actually playing the role of a controlling mechanism for the enjoyment of other rights. In a way it sets the functional limits of all other human rights, especially if the environment is defined in a broader sense, as we believe it should be. In short, if any human activity impairs the environment beyond a certain limit, it should be regarded as an activity producing negative results and be forbidden or changed until it ceases to produce such detrimental results. On the other hand, if human activities do not impair, or if they improve the environment, they should be regarded as beneficial.

INSTITUTIONAL MACHINERY

Efforts should be continued, especially within the United Nations system, for developing a more efficient machinery for implementation of human rights, on a universal as well as on a regional, subregional, and bilateral basis. That means that the United Nations should be more oriented to practical action, even before any more serious UN reorganization is undertaken. This does not mean, however, "that the legislative process in the field of human rights should be halted, if not ended" at this phase, as some authors believe.[17] It seems to us that even some classical human rights could be further developed and more clearly defined.

One of the reasons for this view is the necessity for better protection of human rights from the negative influence of the development of science and technology. Another reason, closely related to the first, is the impact of the environmental problématique. A third is its essentiality for the whole field of development. Besides, further development of the newly emerged human rights, for instance the rights to adequate environment and to development, certainly requires not only that they be more precisely defined but that they be legally regulated in a more concrete and efficient way. States should undertake a solemn obligation to take part individually and through international co-operation in promoting specific programmes and all other appropriate measures in the field of human rights, having in mind the use of science and technology for further protection and promotion of human rights, the environmental issue, and development in economic and other fields.

THE ROLE OF NON-BINDING DOCUMENTS

International documents of a non-binding nature play an important role in all fields of international relations, and the human rights problématique is not an exception. In many cases they can regulate international relations in a specific field de facto, although they are not formally binding. Besides, they often lead to a higher level of regulation, i.e. international treaties, institutional arrangements, etc. The International Declaration on Human Rights was followed, for instance, by the Covenants, and similar developments occurred in other fields. From that point of view the proposed Declaration on Human Rights and Scientific and Technological Developments[18] could be an important step towards better protection and promotion of human rights. It seems to us that such a declaration should contain, more or less, what has already been proposed by the group of experts convened by the United Nations in Geneva in September 1975, but that an additional effort should be made to link it more closely to development and the environment.

Naturally, one must always have in mind the other side of the coin. More so-called soft law does not automatically mean more regulation and further progress in the field. In some cases, it leads to a proliferation of documents with very little or no importance at all, and it could even hinder the process of legal regulation. In other cases, however, including the fields of science, technology, the environment, and some other cognate domains, non-binding norms can be, if functionally well designed, very important and almost universally accepted and applied in such a way as to influence human rights beneficially. Good examples of such norms are formally non-binding ecological standards,[19] which are accepted by all or most interested states and other subjects of international law because it is in their interest to do so. The sanction for those who do not apply the standard becomes "functional." That means that the mere fact of not applying them can cause impairment of the environment, loss of profit, health problems, or even loss of life, lessening of political prestige, etc. If adequately set, they could represent an optimal mode of behaviour. In that way, although formally non–binding, they contribute to the protection of the environment, having a directly or in-directly positive impact on the protection and promotion of human rights.

MILITARY TECHNOLOGIES

Although this paper is not devoted to the military aspect of the problématique, a few words should be said. There is no doubt that the advancement of military technologies is one of the greatest dangers to all human rights, including the most basic right to life itself. The development of nuclear weapons, and other arms for mass destruction, as well as military technologies in general, poses a very serious threat to the environment and *ipso facto* to human rights, even if not used at all in war (radiation caused by nuclear tests, pollution produced by the military-industrial complex, impairment of the environment originating from military manoeuvres, etc). In the case of a large-scale use of such weapons, extinction of humankind as well as destruction of the environment, including all or most of life on planet earth, is possible. Production of various other kinds of sophisticated weaponry is in many cases meant more for domestic use (i.e. various kinds of oppression). Introduction of new technologies, be they new weapons or other police devices, used against criminals who by their illegal conduct are impairing the human rights of other members of society, is justified if based on legal proce-dures and obedience to democratic laws by police and other state services. It is, however, certainly not acceptable wherever such means are used for oppression, based on discrimination of any kind mentioned in Article 2 of both Covenants.

It would be ideal if we could achieve general and complete disarmament and use at least a part of the financial resources thus released "for comprehensive development, in particular that of the developing countries."[20] The realities of our time are not in favour of such an ideal solution. General and complete dis-armament is still not a thing of the near future, but let us hope that the efforts which are being made will bring at least partial success, more efficient interna-tional control and better relations among the two opposing military blocs, as well as the end of various local armed conflicts. Otherwise, especially where nuclear and other weapons for mass destruction are concerned, our future, including further action for protection and promotion of human rights, will not be very bright.

It is interesting to mention, however, that the development of science and tech-nology can help to a certain extent even before total disarmament is attained. For instance, the "technical progress over the past two decades has reduced the pros-pects of purely accidental war."[21] The same progress, and especially the use of space technology for remote sensing of the earth, is already helping superpowers to control each other, and it will make possible, combined with inspections and other means, the control of any future disarmament plan. Finally, it is interesting to add that in some cases improvement of technology can, even in war, prevent unnecessary human losses and suffering. Joseph S. Nye Jr believes, for example, that the increase of accuracy with which weapons can be delivered "could help to reduce reliance on battlefield nuclear weapons and on city-burning strategic weapons."[22]

ULTRA-HAZARDOUS ACTIVITIES

Another important issue covers all potentially very dangerous activities, usually called ultra-hazardous. They could cause (and in many cases have already caused) irreparable damage to the environment. They can cause even a global catastrophe, not only for reasons of technical inadequacy (technical failure of systems which are not well tested, not developed enough, hastily put in use, built without adequate security, etc.), but because of purely human factors like psychic stress, other psychic pathologies, and misperception for different reasons. Even accidents, like the recent disaster in Chernobyl, can give rise to these catastrophes. Very dangerous technologies, as well as scientific experiments, especially those done *in vivo*, should be legally, morally, and by all other means carefully controlled and, when necessary, forbidden. The faster the development of science and technology, the longer will need to be the list of the strictly prohibited activities.

While it is possible to control and to forbid ultra-hazardous technologies and other activities, the handling of the relevant knowledge itself is quite another question. It is evidently not easy nowadays to control much potentially dangerous knowledge, including the results of laboratory experiments, such as those done *in vitro*. Such knowledge, once attained, cannot of course be abolished, and is relatively easy, given place, means, and other circumstances (financial and otherwise) to apply in practice. This means that it could be relatively easily transformed into various means and techniques that could be used for military and other purposes hostile to the environment.

There are many questions in connection with this problem which are not easy to answer. Where, for example, does the freedom of scientific research end? Are all experiments *in vitro* acceptable? In short, where is the red line forbidding further tinkering with nature? It would be impossible, for instance, "to abolish nuclear knowledge without burning all books and all scientists. The prospect for that solution may have passed when the Pope failed to burn Galileo."[23]

DUTY OF DISCLOSURE

One of the possibilities is to lay greater stress on the duty of both states (i.e. through public authorities acting on their behalf) and scientific researchers themselves "to contribute to the definition of the aims and objectives of the programmes in which they are engaged and to the determination of the methods to be adopted which should be humanely, socially, and ecologically responsible."[24] A very important issue in connection with this is the scientist's right to publish research results, including all results relevant to the state of the environment. People have the right to know the real state of the environment,[25] and any attempt of state authorities, industrial enterprises, research institutions, or anybody else to keep vital information secret, or not to disclose it completely, is without doubt a violation of human rights. It is especially so when the information is of crucial importance for human existence.[26]

PROBLEMS OF DEVELOPING COUNTRIES

When it comes to the assessment and control of the development of science and technology, as well as environmental protection, developing countries are in a much worse position than the developed ones. First of all, as a rule, they do not have sufficient numbers of qualified scientists and other skilled personnel, and secondly they lack financial means and laboratory and other equipment. Besides, they do not co-operate sufficiently among themselves, in spite of all declarations to the contrary. In many developing countries the dangers arising from scientific and technological development are not per se regarded as a serious enough problem. Development is sometimes taken as a priority and an excuse for the massive impairment of the environment and violations of human rights. Many mistakes are made, including some of those already made by the developed countries long ago. Weak economies, unstable political systems, military adventures, non-existent or bad environmental policies, unfavourable climate, the population "explosion," and many other factors only make the situation worse.

The developed countries, instead of increasing their efforts to help the developing countries in all possible ways to overcome their hardships and to make our world a better place for all, are often more interested in their own short-term interests. While giving some aid and financial assistance directly or through international organizations, they allow their firms, and especially some multinational corporations, to use the labour and natural resources of the third world without taking adequate measures for environmental protection. The work of the corporations, at least some of them, often impairs the environment even in the industrialized countries themselves. Their activities could be, as Capra wrote, "altogether disastrous in the third world. In those countries, where legal restrictions are often non-existent or impossible to enforce, the exploitation of people and of their land has reached extreme proportions."[27]

It has been repeatedly stressed in many important international documents,[28] conferences, and other fora, and by distinguished experts, that major changes in international economic and other closely interrelated domains are needed. The World Commission on Environment and Development believes, for instance, that: "Two conditions must be satisfied before international economic exchanges can become beneficial for all involved. The sustainability of the ecosystem on

which the global economy depends must be guaranteed. And the economic partners must be satisfied that the basis of exchange is equitable; relationships that are
unequal and based on dominance of one kind or another are not a sound and
durable basis for interdependence. For many developing countries neither condition is met."[29] Both of the conditions should constitute basic elements of any new
international economic order[30] or any other serious attempt to change international economic relations in general so as to better and improve the position of
developing countries in particular. The aim of this article is not to review various
proposals for the introduction of a new international economic order, or other
ideas for improving the position of developing countries, but primarily to stress
the importance of the ecological issue as a basic prerequisite for both the improvement of the economic situation in the world and for the better protection and
promotion of human rights.

DISADVANTAGED GROUPS

Certain groups (ethnic minorities, old people, children, people with impaired
health, women, the poorest strata of society, etc.) are often endangered more
than the rest of the population and therefore deserve special attention. Their
rights are violated in many ways, sometimes by subtle means that are hard to
detect, sometimes openly and ruthlessly. Not having sufficient financial, political, and other power, and being in some ways already underprivileged and discri-
minated against, members of the minority groups are, for instance, forced to live
in the most polluted parts of cities, are less educated, cannot buy good food and
usually consume various kinds of cheap industrially prefabricated "junk food."
They are often addiction-prone (narcotics, alcohol, tobacco), are not able to take
various steps for countering the negative effects of the development of science and
technology (by consuming health foods, by better health care, and in many other
usually costly ways), and are not able to protect their rights by legal means, even
when such means exist in their society. It is evident, however, that legal means
are not sufficient seriously to improve their lot and that other ways and means
must be sought, from better education to better health and other services. Probably more important than all these is improvement of their economic, social, and
political position in society.

TECHNOLOGY ASSESSMENT

It is evident that one of the basic tasks in the future regulation of this problématique should be further development of appropriate machinery, on both international and national levels, for the assessment of the development of science and
technology. That means an approach which is as democratic and, at the same
time, as professional as possible. This is not easy to achieve. It is for governments

to develop appropriate mechanisms for this assessment (taking into account their domestic, legal, political, and general situation). Governments, especially those of developed countries, are trying to do this by resorting increasingly to the engagement of science and technology advisers. Such advisers already play an important role, even in the foreign policy decision-making processes.[31] At the same time traditional politicians and diplomats play an important role in the decision-making process on both the national and international scenes. The engagement of experts does not automatically solve all problems: in some cases the experts, as W.J.M. Mackenzie writes, "speak in fact for vested interests, or they may be totally at a loss and in hopeless disagreement."[32] Politicians, for their part, are not, as a rule, expert in science and technology, and are influenced, among other things, by "the demands of political survival."[33] All these factors, and many others, play a role in final decision-making in the field of science and technology.

Decisions concerning policy in the fields mentioned are at present usually made in the closed circles of various government offices, other official institutions, formal and informal groups and cliques, large industrial enterprises, etc. In most cases they are interested primarily in some kind of profit (financial, political, professional, etc.) while the interests of others, especially in the long term (other companies, other countries, population as a whole, mankind, future generations, etc.) are low on their priority lists. Technical élites,[34] closely linked with military, industrial, political, and other circles of power, are often reluctant to disclose their activities. Besides that, the very nature of the development in most fields of science and technology prevents most people even from understanding what is going on. Who is to control scientists and technicians in their highly specialized fields? The old Roman maxim, *Quis custodiet ipsos custodes*, becomes more than ever real. It is evident that further democratization of the decision- and policy-making processes is the best means of, on the one hard, protecting human rights from the negative influences which could flow from the development of science and technology, and, on the other, increasing the beneficial impact of this development.

THE RIGHT TO PARTICIPATE

The right to participate, including the right to be informed (to seek and receive information), in all activities concerning the development of science and technology, as well as economic development, environmental protection, and other cognate fields, is of the utmost importance. "The voluntary and democratic involvement of people in contributing to the development effort"[35] by, among other things, taking part in decision-making, formulating policies, and controlling various activities, including those in the fields discussed in this article, is one of the ways to protect human rights from the negative influence of all kinds of deviations in policy-making. Various forms of participation, self-management systems,[36] referendums, special parliamentary commissions, public opinion, the mass media (and especially a concerned and inquiring press), professional associations (with their codes of professional ethics), ecological, pacifist, and other movements, consumer associations, as well as other formal and informal groups and individuals (including scientists in their role as conscientious people acting independently of the establishment), are only some of the possible ways of making the decision-making process as open and as democratic as possible, and of creating "a beneficial framework for individual and group self-realization, as well as for the healthy connection between state and society."[37]

There are many problems in connection with the democratization of the decision- and policy-making process. It is true, especially in developed countries, that, generally speaking, "people are sufficiently well educated" to take part in the process of defining goals of scientific and technological development.[38] In many cases, however, it is not easy to find experts impartial enough to judge and, if necessary, participate in the reassessment of a potentially dangerous technology or hazardous scientific experiment. This is especially the case in most developing countries, in which there are not enough experts even to properly assess imported foreign technology or chemicals, or to give advice on the extent to which an investment in industry would be environmentally sound. In fact, most developing, and in some fields even a number of developed countries, must rely on foreign expertise if they want to assess a technology or product, and since it is in the financial interest of foreign exporters – who come usually from developed but also from some developing countries – to show their otherwise polluting technology or unhealthy product in the best possible light, it happens that the experts asked to give their judgement are not always unbiased or entirely free of the possibility of bribery. One of the roles of international organizations, especially those belonging to the UN system, is to take part in the process of assessment, as well as to provide interested countries on request with all needed expertise for their development in science and technology and for environmental protection, in an impartial and professional way.

HUMANITARIAN LAW

The role of humanitarian law, as another autonomous branch of international law (much older, incidentally, than human rights law), should be given due regard. We share the opinion that "the links between human rights and humanitarian law are real and growing stronger."[39] In certain circumstances these two complementary, although still distinct, systems become "convergent and perhaps interpenetrating."[40] It is a fact that the International Committee of the Red Cross and the national societies of the Red Cross and Red Crescent, as well as the Henry Dunant Institute, are increasingly active in the field of human rights, not only in armed conflict but in areas such as health, education, the protection of victims of natural disasters, and the protection of the environment.

It must be noted that serious disasters caused by human activities could be, under certain circumstances, regarded as a "public emergency which threatens the life of the nation," recognized in the International Covenant on Civil and Political Rights (Article 4). Cases like the Chernobyl accident, and other serious industrial accidents which have occurred in recent years, show that such a possibility is unfortunately not merely theoretical. Certain derogations of human rights "strictly required by the exigencies of the situation" could be expected, and international law should guarantee that the derogation does not last longer than necessary and that it does not infringe some of the basic (or sacrosanct) human rights. International co-operation should in such cases play an important role in both the protection of victims of the disaster and in giving proper assistance to the stricken country. Humanitarian law and international humanitarian organizations will certainly play an important role in such situations and the process of further convergence between human rights law and humanitarian law will continue.

STRENGTHENING OF AGENCIES

The strengthening and better organization of the existing UN bodies, and especially the Commission on Human Rights and the Human Rights Committee, is needed. The establishment of a permanent institution for assessing the impact of science and technology on human rights remains certainly one of the important steps towards better implementation of the steadily growing body of human rights law in general. It is necessary, however, that such a future institution, if established, devote special attention to development as "a comprehensive economic, social, cultural, and political process,"[41] as well as to environmental protection, which must naturally be done in co-operation with UNEP and other UN bodies.

Regional development deserves special attention, both from theoretical and practical points of view. Various groups of countries (West European, CMEA countries, American countries, etc.), linked by political, economic, territorial, and other ties, have been developing their own practices, as well as a theoretical base in the field of human rights, which certainly deserve further study and, whenever acceptable and desirable, transference to the universal level. Very interesting, indeed, have been developments on the basis of the European Convention for the Protection of Human Rights and Fundamental Freedoms (1950). In spite of the fact that the experiences of the West European countries, which share many political, economic, cultural, ideological, and other similarities, cannot be transferred automatically to others, they at least show that it is possible to improve international co-operation in the field of human rights by using legal means and procedures.

Finally, it is important to add that the experiences of various bilateral bodies, non-governmental organizations, national institutions and international, as well as national, professional associations could also make an important contribution to the development of the protection of human rights, as well as development in other fields, including the environmental issue, which are closely related to the human rights problématique.

IMPROVEMENT OF INTERNATIONAL RELATIONS THROUGH HUMAN RIGHTS CO-OPERATION

More intensive and constructive co-operation is needed in the field of human rights, in science and technology, in environmental protection and in all other cognate fields, as well as general improvement in international relations, as already mentioned. Besides the highly politicized, and without doubt important, questions of disarmament, economic relations, the position of developing countries, armed conflicts in various parts of the world, the situation in South Africa, etc., there are many other possibilities for improvement of international relations in the field of health protection, protection of the environment, and other cognate domains which are only seemingly "of a technical nature." Improvement of co-operation in these areas can make an important contribution in the field of human rights, as well as in other domains, and thus bring about the improvement of the international situation in general.

One of the most interesting and needed areas of possible improvement in international co-operation is comparative research on various aspects of the inequalities in health between different countries. There is a relative shortage of international studies of this kind, even where the developed countries are concerned, because of, among other things, the differences in the standard methods for measuring equality.[42] Standardization of these methods and more co-operation in this important field of research are needed.

There are other interesting cases of international "functional" co-operation[43] which contribute, to a lesser or greater degree, to the protection and promotion of human rights, and which are closely interlinked with both the development of science and technology and environmental protection. Any successful, or relatively successful, international programme of environmental protection – for example, the Mediterranean Action Plan and the activities concerning the implementation of the Convention on Long-Range Transboundary Air Pollution (Geneva, 1979) – not only makes an undoubted contribution to the protection of the environment in the narrow sense but also produces a beneficial influence in other fields, including political relations. Gradual development of international regimes, based on legal and other means, which "govern various dimensions of economic and social interdependence among states,"[44] may reduce the degree of potential conflict and improve the overall co-operation among states.

The vast field of science and technology is certainly the most important of the above-mentioned dimensions of interdependence. The international legal regulation, as well as the regulation by all other acceptable means, of this complex field of activity is of crucial importance not only for international relations, but for the further development, and even survival, of mankind and perhaps all other living creatures on our planet. The closely interlinked human rights and environment problématique is certainly the most important dimension of the process. Without promotion of human rights, from the basic right to life to civil rights and fundamental freedoms and the right to an adequate environment, the development of science and technology cannot rest on firm ground. Only if applied in a humane (and that means, among other things, an ecologically sound) way will it serve its purpose.

INTERLINKAGE OF SUSTAINABLE DEVELOPMENT AND ENVIRONMENTAL ISSUES

The environmental issue undoubtedly adds a new dimension to the problématique of human rights. In the first place, it shows once again that all human rights are closely interlinked, and, secondly, that the problématique of human rights is inseparable from practically all other processes in human society, and especially from economic development and the progress of science and technology. The main conclusion – that the most acceptable model of further development of human society is the model of sustainable development – has its roots primarily in the environmental issue. Policy in all fields of human activity must be environmentally sound. This is especially so in the field of human rights, which cannot be enjoyed without an adequate environment. Further development of science and technology will be beneficial to human society (i.e. it will further promote human rights) only if it is environmentally sound. Besides that, the environmental issue shows in a very clear way that all human rights should be regulated and enjoyed in a balanced way, or to put it better, in a sustainable way. That means that civil and political rights, on the one hand, and economic, social, and cultural rights on the other are needed equally and should be protected and promoted by all means. Finally, the rapid development of environmental law, together with the closely interlinked, and dialectically inseparable, law of sustainable development, are contributing to the development of international law in general and especially human rights law. We fully agree with the statement of His Excellency Judge Nagendra Singh that the efforts of the World Commission on Environment and Development to, *inter alia*, "forge and develop the law governing the environment . . . opens up a new chapter in the history of international law."[45] This is so because that effort succeeded in efficiently combining environment and development in the concept of sustainable development.

Naturally, this ecological and holistic view collides with many present human activities, especially in the field of science and technology. A good deal of what is nowadays regarded as the "progressive development of science and technology" has to be reconsidered and changed or entirely stopped. That means also that a good deal of economic activity must be transformed into what is ecologically sound and socially and politically acceptable. Besides that, many other activities, including life-styles, should be changed, especially in the developed parts of the world. All this, we are aware, is not easy to attain, especially in the short run. The realities of our world will allow only a slow, step-by-step approach but that does not mean that the distant goal of a more ecologically, economically, morally and politically sound, and *ipso facto* a more just, world order should not be sought.

NOTES

1. The beginning of more systematic and organized UN work on the question is usually dated to 1968, or, more precisely, to the International Conference on Human Rights, held in Tehran that year, as well as to General Assembly Res 1450 (XXIII) of the same year. See, for more details, Yo Kubota, chap. 6.
2. V.W. Rasmussen, "The Peril of Ecological Illiteracy: Thoughts for the Graduating Class," *Yale Review*, vol. 75, no. 4 (1986): 594.
3. *Declaration of the United Nations Conference on the Human Environment*, part I, para 1.
4. W.J.M. Mackenzie, *Biological Ideas in Politics* (Penguin Books, London, 1978), pp. 16–17.
5. Mackenzie (note 4 above), pp. 30–31.
6. There are, for example, more than 120 multilateral treaties in the field. See UNEP's *Register of International Treaties and Other Agreements in the Field of the Environment* (UNEP/GC 14/18 and later additions).
7. For example, the Yugoslav Constitution of 1974 (Articles 87, 192, and 193).
8. Article 24.
9. Principle 1.
10. World Commission on Environment and Development, *Our Common Future* (Oxford University Press, 1987), Annex 1.
11. For example, in the Universal Declaration on Human Rights: Article 3 (the right to life): Article 22 (economic, social, and cultural rights); Article 25 ("a standard of living adequate for the health and well-being"), etc. Among other elements of the right, "the improvement of all aspects of environmental and industrial hygiene" is mentioned in Article 12 (para 2b) of the International Covenant on Economic, Social and Cultural Rights.
12. Principle 1 of the Declaration of the United Nations Conference on the Human Environment.
13. Paul Sieghart, *The Lawful Rights of Mankind* (Oxford University Press, Oxford/New York, 1985), p. 167.
14. Sieghart (note 13 above).
15. Preamble of the Declaration, UNGA Res 41/128 (Annex).
16. *Seminar on Human Rights and Scientific and Technological Developments* (New York, 1972), p. 7 (UN ST/TAO/HR/45).
17. Dr T.O. Elias, *Africa and the Development of International law* (A.W. Sijthoff-Leiden Oceana Publications Inc., Dobbs Ferry, N.Y., 1972), p. 213.
18. United Nations, *Human Rights and Scientific and Technological Developments* (UN, New York, 1982), p. 76.
19. Ecological standards can also be formally binding legal norms. For more about ecological standards, see: P. Contini and P. Sand, "Methods to Expedite Environment

Protection: International Ecostandard," *American Journal of International Law*, no. 1 (1972): 37–59; Vid Vukasovic, *Rad Programa UN za covekovu sredinu na njenom medunarodnopravnom regulisanju* (Institute of International Politics and Economics, Belgrade, 1985), pp. 49–63. The general term "ecological standards" applies to all rules, formally binding or non-binding, in the field of environmental protection. In different documents they are, when it comes to terminology, called different things (lists of standards, technical rules, codices, eco standards, etc.).

20. *Declaration on the Right to Development*, UNGA Res 41/128, Article 7.
21. Joseph S. Nye Jr., "Ethics and the Nuclear Future," *The World Today*, vol. 42, nos. 8–9 (1986): 152.
22. Nye (note 21 above), p. 153.
23. Nye (note 21 above), p. 152.
24. See Unesco, Recommendation on the Position of Scientific Researchers, adopted on 20 November 1974 (especially chap. IV).
25. See, for example, *Our Common Future* (note 10 above), p. 330.
26. *Our Common Future* (note 10 above), fn. 24.
27. Fritjof Capra, *The Turning Point: Scientific Society and the Rising Culture* (Fontana, London, 1984), p. 234.
28. Efforts to improve international economic relations are often linked to both the protection of the environment and the protection and promotion of human rights. For example, the UN Declaration on the Right to Development (GA Res 41/128), the World Charter for Nature (GA Res 3201-2/S-VI); the Charter of Economic Rights and Duties of States (GA Res 32/XXIX), etc.
29. *Our Common Future* (note 10 above), p. 67, fn. 25.
30. For NIEO see K. Hossain, ed., *Legal Aspects of the New International Economic Order* (London/New York, 1980); M. Bulajic, D. Pindic, and M. Marinkovic, ed., *The Charter of Economic Rights and Duties of States* (Belgrade, 1986).
31. Ralph Sanders, *International Dynamics of Technology*, Contributions in Political Sciences, no. 87 (Greenwood Press, London, 1983), p. 259.
32. Mackenzie (note 4 above), p. 64.
33. Mackenzie (note 4 above), p. 64.
34. Robert S. Cohen writes that "technical élites have finally come to their own peculiar roles, their power deriving from specialized competence; they are partially insulated from other élites and from democratic decision-making by a scientific and technological sophistication which easily allows for esoteric secrecy (whether military or industrial). *International Social Science Journal*, vol. 34, no. 1 (1982): 69. "Science and Technology in Global Perspective."
35. United Nations, *The United Nations and Human Rights* (UN, New York, 1984), p. 232.
36. The UN General Assembly in 1983 recognized that popular participation, including self-management, "constituted an important factor of socio-economic development, as well as of respect for human rights" (*The United Nations and Human Rights* (note 35 above), p. 233. This important subject was on the agenda of the UN General Assembly, the Economic and Social Council, and the Commission on Human Rights, as well as being the main topic of international conferences (for instance, the International Seminar on Popular Participation held in Ljubljana, Yugoslavia, 17–25 May 1982). See also P. Jambrek, "Participation as a Human Right and as a Means for Exercise of Human Rights" (Unesco Division of Human Rights and Peace, 1982) (Unesco doc. SS-82/WS/54).

37. Richard Falk, "Nuclear Weapons and the Renewal of Democracy," *Nuclear Weapons and the Future of Humanity* (Rowman & Allaheld, Totowa, N.J., 1986), p. 439.
38. Hajime Eto and Ryujiro Ishida, "Integrating Assessment in National Technological Policy," in Jacques Richardson, ed., *Integrated Technology Transfer* (Lomond Books, 1979), p. 121.
39. A.H. Robertson, *Humanitarian Law and Human Rights*, studies and essays on international humanitarian law and Red Cross principles in honour of Jean Pictet (Geneva/The Hague, 1984), p. 800.
40. Robertson (note 39 above), p. 802.
41. Preamble of the UN Declaration on the Right to Development, UNGA Res 41/128.
42. Julian Le Grand writes, for instance, that: "In the case of health the conventional procedure has been to compare the health (usually mortality) experience of different social or occupational classes . . . Any attempt to apply this procedure to international comparisons, however, encounters the obvious difficulty that definitions of social class, and of the occupational classifications that underlie them, vary widely from country to country. The fact that the only successful comparisons have been undertaken between countries with great similarities in culture (and therefore in occupational classifications) reinforces the point." J. Le Grand, "Inequalities in Health – Some International Comparisons," Papers and Proceedings of the First Annual Congress of the European Economic Association, 29–31 August 1986, Vienna; *European Economic Review*, vol. 31, nos. 1/2 (1987): 183.
43. Winfried Lang, "Environmental Protection – The Challenge for International Law," *Journal of World Trade Law*, vol. 20, no. 5 (1986): 495.
44. Nye (note 21 above), p. 154.
45. Nagendra Singh, "Sustainable Development as a Principle of International Law," inaugural address at the seminar convened by the Free University of Amsterdam, 9 April 1987, p. 21.

BIBLIOGRAPHY

Rybczynski, Witold. *Taming the Tiger – The Struggle to Control Technology*. Viking Press, New York, 1983.
Sieghart, Paul. *The Lawful Rights of Mankind*: An Introduction to the International Legal Code of Human Rights. Oxford University Press, Oxford/New York, 1985.
United Nations. *Human Rights and Scientific and Technological Developments*. United Nations, New York, 1982.
——. *The United Nations and Human Rights*. United Nations, New York, 1984.
World Commission on Environment and Development. *Our Common Future*. Oxford University Press, Oxford/New York, 1987.

Part 5

10

Conclusions and Recommendations

C.G. WEERAMANTRY

This volume draws together many strands of thought regarding the affirmative role of technology in a world order which will focus increasing attention on human rights.

As we move towards the close of this century and as the next gets into its stride, old concepts of sovereignty will become less rigid. The watertight barriers that insulated nations will become more porous. This gradual transformation of the concept of sovereignty can well be one of the key features of future development in the spheres of international relations and international law.

The universal nature of technology will make it one of the key factors in breaking through state boundaries, for technology does not recognize national frontiers in its multifarious impacts upon life and society. Technology, which will thus play a great role in promoting universalism, will also have other important roles to discharge in making life better for all individuals in that more universalist society. It will at the same time be a powerful catalyst on the international scene and at the level of individual rights, and its role can be both a positive and a negative one. Either way its impact will be extremely powerful, with its capacity for influencing human rights at every level from the micro-level of the individual to the macro-level of international relations.

Since technology will play this key role in making our planet one world and in reducing the barriers that have thus far impeded this concept, it is time for the world community to give thought to directing technology on a course which will maximize its use as an instrument for the furtherance of universally accepted norms of human rights. This study and others have made it clear that science and technology are forces that are now too powerful and too full of ramifications for them to be left to laissez-faire attitudes which permit them to take what direction they please. This study makes it clear that there is much scope for the process of guiding and channelling technology so as to make it an instrument for the furtherance of human rights in the developing world.

The multiple choices of direction that open out with each fresh advance in science make the process of selection a particularly significant one. The developing world, from being a passive recipient of technology tailored to suit other societies, must take a guiding hand in influencing the fashioning of technologies that suit its own needs.

The problem before us is vast and only some selected perspectives can be presented in a single volume such as this. The course we have chosen to follow is the threefold approach of presenting global perspectives, international responses to the problem, and particular studies on some specific problems.

In the first category is the conceptual and theoretical study by Dr Herrera. This stresses the prospective view and the role of normative prospective studies. Dr Herrera emphasizes the need for considering the problems involved in an interdisciplinary context, rather than, as is often currently the case, in terms of a single component such as economics, and asking for the support of other disciplines afterwards. He also raises the question of the need to steer ourselves away from the danger of falling into the "defensive" approach of identifying the possible negative impacts of new technologies. The potential of the new technologies should be taken into account affirmatively in the establishment of socio-economic objectives.

Dr Herrera provides much food for thought in pointing out that in this field, as in others, we must break out of the fashion of regarding the developing world as a dependent variable of what will happen in the advanced countries. In Dr Herrera's language, "a wide array of new options is opened and they offer the third-world countries the opportunity to participate actively in the construction of a new and more equitable world order."

Forecasting is not simply a theoretical exercise but must always be performed as a guide for action. The methodology used for the TPLA Project (Technological Perspective for Latin America) to which Dr Herrera refers us gives us many insights into the sequence of steps involved in prospective studies, as well as in regard to the criteria of the desirable society which we postulate as our goal. Science and technology should be explicit variables incorporated in the whole integrated process of socio-economic planning, which should take the place of the present dichotomy between socio-economic and R&D planning.

An important conceptual challenge thrown out in this chapter is the need to work towards the formulation and elaboration of the right of participation in all social decisions and of the right of access to intellectually creative work.

An important aspect to be borne in mind in addressing the question of turning technology to the furtherance of human rights is Dr Herrera's observation that our aim should be to close the technological gap not in absolute terms but in the context of the required socio-economic and institutional adaptations. Too often we are led astray from the practical issue at hand by the insurmountable difficulties of striving to close this gap in absolute terms. The apparent impossibility of that task should not prevent us from attaining immensely valuable results through adaptations resulting from thoughtful interdisciplinary studies.

Dr Herrera's analysis, although presented as a view from the third world, has a broader validity which makes it relevant to all societies.

Dr Chamarik pursues this scheme more specifically in relation to the third world, emphasizing the concept of self-reliance in science and technology. The most pertinent sphere for the application of this principle in relation to developing societies is that of agriculture.

Dr Chamarik steers us away from reliance on technological decisions taken for third-world countries by élitist groups in those countries. The needs and wishes of those élitist groups may often be at variance with those of the grass roots, and if one is thinking of furthering human rights in a manner meaningful to the bulk of the people in those societies, such élitist decision-making is not the answer. It leads to the pursuit of such aims as accelerated economic development towards industrialization. Such an approach can lead to the pitfalls of dependence and subordination.

Agricultural societies have never been without technological knowledge and inventiveness. Their traditional means of learning and skills in technological adaptation and innovation need not be minimized or ignored, for they are directly related to a people's real and relevant needs and environmental conditions.

From the standpoint of human and social progress, the modern and traditional must be looked upon as complementary to each other.

Conditions of self-reliant development are pursued through the Self-reliance Study of six Asian countries conducted under the aegis of the United Nations University. This leads to a shared development perspective rather than a solution which seeks self-sufficiency for each country exclusive of all others.

The various ingredients of technological self-reliance are interrelated. These involve principally an optimal use of local resources, the development of indigenous human resources and the development of grass-roots institutions.

Dr Chamarik concludes that the real solution lies in the technological and productive capability within the rural communities themselves. The choice and assessment of technology, instead of being imposed or forced upon them, must be made by and within the rural communities themselves.

An important feature of Dr Chamarik's analysis is that the socio-cultural factors necessary for technological self-reliance will be impossible to achieve without favourable politico-economic conditions. Lack of autonomy in the decision-making process, which has been forced upon many traditional Asian societies, needs to be revised, after centuries of dependence and underdevelopment, through a process of revitalization in the politico-economic sphere.

There is no inherent incompatibility between modern and traditional technology. The path of future development can be changed for the better by human intervention. Modern technology, in a symbiotic relationship with the traditional, must supplement rather than supplant indigenous technology. In short, the question is not whether Western science should be made use of, but how, on what conditions, and with what objectives.

The third essay in the series of global perspectives is Dr Farer's examination of human rights and scientific and technological progress from a Western perspective.

Dr Farer supplies a corrective to any tendency to think of technology and human rights only in terms of technologies that produce obvious impacts upon group-oriented third-world ways of life. As Dr Farer observes, other studies in this volume, which approach problems of technology and human rights from the standpoint of the third world, do not discount the importance of individual rights and individual autonomy.

One cannot lose sight either of the fact that within third-world societies there are considerable groups whose attitudes and values run very close to those of the individually oriented approaches of the West. Dr Chamarik, for example, has dwelt at some length on the dominance of such groups within developing societies.

Since there is an interrelatedness of human rights which does not permit the isolation of one group from another, it is important to our sense of perspective to consider the impact on human rights of some technologies which may apparently have relevance only to societies of the West.

There is no way of eliminating a concern with those technologies in reviewing the impact of technology on any section of the world's population. Technology knows no boundaries of nationality, class, or creed, and sooner or later a technology which is thought to be pertinent only to one section of the global population makes its impact upon another, however sophisticated or exclusive that technology may appear to be.

Dr Farer selects three areas of technology for special consideration – nuclear weapons, procreation and child-rearing, and privacy.

All of these, though apparently far from the concerns of the developing world, affect it intensely. In relation to nuclear weapons there is no question that they tear the seamless fabric of universal human rights so grievously as to affect the basic human rights of all people everywhere. We cannot talk meaningfully of using technology for the service of third-world peoples without addressing the problem of the conceptual framework underlying the philosophy that nuclear weapons are permissible in any circumstances. Unless that philosophy can be countered it is academic to address any other questions of using science and technology in the service of human rights.

The first problem Dr Farer addresses is therefore as much a problem of the developing world as of any other segment of the world community.

Procreation and child-rearing by their very nature impinge intimately on the humanness and dignity of every human being. If there are potentially dangerous uses they are potentially dangerous everywhere, as for example one of the unorthodox uses Dr Farer mentions, namely their use in imaginable totalitarian political and social settings. If there are potentially beneficial uses, they are potentially beneficial everywhere and one would be rash to postulate any time interval between their uses in developing and developed societies. Moreover, even if they do not have a sense of immediacy for third-world societies, they have an immediate impact upon the concept of human personality.

The third technology, that which bears on privacy, has obvious importance for all societies, developing and developed alike, and indeed their potential uses for the former are far-reaching indeed. The capacity to deliver information from a source located anywhere on the globe to receivers in any other part of the globe gives this technology the most far-reaching potential for service.

Dr Farer's article demonstrates also the inaccuracy of right-wing Western polemical attempts to argue that a fundamental difference exists over the centrality of human freedom between North and South and that the discourse of human rights is used by the latter for effecting a transfer of resources from North to South.

These global overviews conclude the second part of this volume. We move from there to the international response, the subject of part 3.

We have moved a long way from the early UN attempts to harness science and technology for the benefit of the developing world.

The initial attempt, the UN's World Conference on the Application of Science and Technology in Less Developed Areas (UNCSAT), held in Geneva from 4 to 20 February 1963, brought together 1,665 participants from 96 countries, as well as a multitude of international organizations. However, to quote the words of one commentator (Professor Volker Rittberger in *Development and Co-operation* (German Foundation for International Development, vol. 3 (1979): 14),

This world conference resembled more a global science and technology market than a consulting and negotiating forum to promote concrete development programmes. Accordingly this world conference – in contrast to the world conferences of the seventies – neither passed a declaration of intent nor an action programme. Demands of individual developing countries to create a new organization within the system of the UN which would specialize in the promotion of the application of science and technology for development purposes fell on deaf ears.

In the quarter-century that has elapsed there has been much serious consideration of the problem from both theoretical and practical standpoints. While developmental aspects have loomed large, it is seen that development is only one segment of the much broader frame of human rights within which the problem must be considered. Various action programmes have been set up and institutional mechanisms set in place. In 1971 the Advisory Council for the Application of Science and Technology to Development (ACAST) published a "World Plan of Action for the Application of Science and Technology to Development." Numerous international declarations have ensued. At the thirty-first session of the General Assembly in 1976, a formal resolution was passed calling for the establishment of a United Nations Committee on Science and Technology for Development (UNCSTAD). All these topics are the subject of continuing discussion and ongoing action programmes.

Dr Yamane's chapter on the Normative Response of the International Community provides an essential conceptual background to this study by tracing the evolution of the instruments which have been fashioned to cope with human rights problems caused by scientific and technological progress. As a prelude to this Dr Yamane goes through the important process of a stocktaking of the human rights which are particularly in danger of erosion by scientific and technological progress.

The catalogue of rights that might be directly affected is considerable, and when one adds to this a list of rights that might be adversely influenced, the list is such as to cause concern, for a large section of the spectrum of human rights is under direct or potential attack.

The international instruments which are a means of protection against the abuse of science and technology are both general and specific. Specific instruments relate to the right to life, the right to physical and spiritual integrity, the right to privacy, and the right to information.

Dr Yamane also lists the instruments fashioned as a means to assure positive uses of scientific and technological progress for advancing human rights, dividing them again into instruments of a general and a specific character. In the latter category are those which cover the right to benefit from scientific progress and the right to an adequate standard of living.

Dr Yamane points out that the major preoccupation of the United Nations in dealing with the impact of science and technology on human rights has turned to economic development, coinciding with the ongoing effort of the international community to establish a new international economic order. The Charter of Economic Rights and Duties of States of 1974 confers important rights and imposes important duties upon states, both developing and developed. An important duty is the duty of the developed countries to co-operate with the developing countries in the evolution of scientific and technological infrastructures and other activities so as to expand the economies of the latter group.

Developments such as these open up areas for important conceptual debate. In particular, calls for the active intervention of the state in many spheres of life clash with the liberal-individualistic approach to human rights. Dr Yamane rather laments the effort spent on superficial conceptual compromises rather than on devising effective ways and means of implementing human rights. She points out that at the same time further conceptual fuzziness is introduced by the indiscriminate proclaiming of rights of all kinds. Perhaps a way out of this impasse is, she suggests, a specification of the areas for implementing different categories of human rights and a dissociation from the broad general issues of economic development such as technology transfer and exploitation of natural resources.

This call for a sharpness of focus in our thinking about human rights is timely, for overenthusiasm can damage the entire cause by blurring conceptual clarity. We need to develop human rights and at the same time preserve their conceptual integrity. Expansion of categories which is too rapid or indiscriminate can undermine the foundations of the basic conceptual structure.

Conceptual development in the field of human rights is shown by Dr Yamane to be haphazard. One is reminded of the way in which the common law system was built up through "a wilderness of single instances" rather than through a cohesive body of unifying principles. Later, the text writers got to work on this unwieldy mass of material and reduced it to some form of coherence and logical consistency. Perhaps the field of human rights awaits such pioneering academic labour. Perhaps, as Dr Yamane suggests, the crucial issue centres on the control of information, for, without information, considered decisions on matters of science and technology are impossible. Her call for the elaboration of an international instrument which will reinforce the right to information could be a major conceptual outcome of this study.

With the benefit of the conceptual insights resulting from Dr Yamane's study we pass on to Dr Kubota's careful analysis of the institutional response.

Having traced the steps leading, especially through Japanese and Yugoslav initiatives, to the present project, he points out that the question of the effect of scientific and technological developments on human rights was not considered in detail until the International Conference on Human Rights in 1968. This resulted in the Proclamation of Tehran. The General Assembly by its resolution 2450(xviii) of 19 December 1968 invited the Secretary-General to undertake detailed studies in such areas as privacy in the light of advances in recording technologies and the protection of the human personality in the light of advances in biology, medicine, and biochemistry.

From that initial stage Dr Kubota takes us in detail through the developments, studies, and reports that emerged during the period 1971–1987. A landmark event was the General Assembly's 1975 Declaration on the Use of Scientific and Technological Progress in the Interests of Peace and for the Benefit of Mankind, which the Commission on Human Rights, in its resolution 10B(XXXIII), described as a guide for its future work. The Commission on Human Rights also emphasized that states should take account of the provisions and principles contained in that Declaration. Major reports prepared by the Secretary-General and the specialized agencies concerned are carefully noted. The growing volume of UN documentation has reached such proportions that some guidance is required through this mass of material, and Dr Kubota's chapter provides the researcher with the assurance that he has not lost sight of significant documentation in this field.

Dr Kubota, while not denying the importance of the work undertaken to date, emphasizes the need for guidelines and rules of conduct in scientific and technological activities. These guidelines should preserve "the balance which should be established between scientific and technological progress and the intellectual, spiritual, cultural, and moral advancement of humanity."

The need for information emerges strongly from Dr Kubota's article as it does from some of the other contributions in this book. This is because there is a need for everyone to defend himself or herself against the negative effects of science and technology, and to do so one needs access to information on potential dangers. Dr Kubota therefore sees freedom of information as lying at the heart of the protection and promotion of human rights.

The decision-making process, the appellate process, the legislative process, and the administrative process all require this free flow of information. Dr Kubota notes in this regard the efforts of the United Nations Educational, Scientific, and Cultural Organization and the International Social Science Council in this field. These are important pioneering efforts which need to be more widely known.

Dr Kubota argues that policies, approaches, and attitudes which create or maintain an artificial dichotomy between legal and human rights considerations and scientific and technological activities should be abandoned or rectified. Such "artificial compartmentalization" no doubt colours the thinking of many decision-makers. Dr Kubota sees the establishing of new guidelines as a means of facilitating the transfer of scientific and technological activities from the purely technological and scientific arena into the legal and political arena.

This is important work and is in line with the series of projects of the United Nations University on scientific and technological developments, which have the common objective of setting forth guidelines enabling scientific and technological policies to make a maximum contribution to human and social development needs. The current project, as Dr Kubota points out, represents the first occasion on which the United Nations University has decided to respond to an invitation by a UN human rights body to launch a project of this kind. It is therefore specially important that it should bring about new and fruitful achievements in this field.

Part 4 deals with specific problems, of which the structure of the scientific enterprise, development and human rights, and the environmental problem have been selected as three important areas worthy of special attention.

Any study of the ways of turning science and technology to public advantage needs to address the question of the structure of the scientific enterprise, for scientific decision-making cannot be undertaken without this knowledge. The scientific enterprise has a logic and a scheme of its own.

Dr Nakayama undertakes this study in his contribution to this volume. By taking us through the various stages in the production of industrial science and defence science he assists us in focusing attention on areas of secrecy and impenetrability. Much though we may desire to make the human rights influence felt in these areas of decision-making, there are very real obstacles stemming from considerations of corporate profit or strategic secrecy, which make the social assessment of science difficult at many stages of production. Assessment is difficult enough. Making an impact on the processes of decision is even more difficult.

The theoretical openness of scientific knowledge and its theoretically universal nature thus tend to be negated across a considerable spectrum of the scientific enterprise. This amounts in effect to a denial to scientists of basic human rights of dissemination and discussion of scientific information. The impact on scientists' rights is more far-reaching than this, for it touches human rights in other ways as well, including questions of hazards to which scientists are exposed in consequence of such secrecy. The public too are the sufferers, for the damaging effect upon the environment of certain scientific processes is not known to the public owing to such secrecy, and hence is not the subject of preventive action.

Service science was to a large extent free of many of these obstacles, but many aspects of service science are being industrialized now. Even in such an area of medical service as organ transplants, the production of artificial organs is rapidly becoming big business.

All of this raises significant conceptual problems. Dr Nakayama brings up the important question of the disadvantaged position in scientific decision-making of those who lack the necessary knowledge – and that goes for most of us. Do we have a right not to be at a disadvantage because of our ignorance?

The answer currently tends to be along the lines that ignorance must be cured by information. If people continue to remain ignorant they must pay the price for this. But the proliferation of knowledge is occurring on such a scale that it is idle to postulate a duty to acquire knowledge in every area of technology that may impinge on our lives. The degree to which individuals are disadvantaged by ignorance will grow in every society, and especially in the conditions of the developing world the danger is already acute.

Inaccessibility to information and maldistribution of information are vital areas needing attention if technology is to be turned in the direction of service to the community. Information pollution, as Dr Nakayama terms it, is a real danger present in every society. It is a growing danger everywhere and in present world conditions is one of the acutest problems that third-world populations face.

Human Rights, Technology, and Development is the second selected topic in part 4. This essay, written by the editor, seeks to explore the impact of technology in the light of the three principal elements involved in the General Assembly's Resolution of 1984 on the right to development: participation, contribution, and enjoyment.

In order to increase participation in scientific decision-making one has to explore the decisional routes through which new technology enters a developing country. The decision is often taken under pressures external to the developing country in question, and research is needed to ascertain with more precision the factors influencing decisions and the points at which such decisions are taken.

Such considerations point again to the need for more diffusion of information regarding both the decisional factors involved and the alternative technologies available.

Technology surveillance, early recognition and alert systems, international exchanges of technological assessment, and technical co-operation among developing countries all need to be stepped up. An important part of the next stage of research upon the current project may well be an investigation of the means by which such institutional mechanisms can be set up. Whether such centres are set up nationally or regionally, there needs to be a network of them spanning the entire spectrum of developing countries. Some countries may lack the necessary resources for all these purposes, but the paucity of their resources is no reason for the denial to their citizens of their undoubted human rights, which will be impaired by the lack of such mechanisms. This is therefore an important area for regional and international co-operation, for if human rights are involved – as this volume seeks to demonstrate – there can be no excuse for the global community's neglect of its obligation to assist such countries in this essential task.

As with decision-making, so also with contribution to the technology itself, the scope for third-world participation must be increased. We need research to identify the optimum points for such input and to maximize their availability. Participation in product designing, the multifarious stages involved in joint routine enterprises and the legal considerations associated therewith, and education for participation, as well as appropriate technology, need to be examined from this point of view. The formal network for the distribution of information does not reach the grass roots, and other methods must be explored.

The third leg of the tripod – enjoyment – is perhaps the most important. Much technology which is suitable does not reach the bulk of the people who should be enjoying it. This is indeed the greatest challenge in the process of making technology serve the purpose of human rights. It necessitates an examination, among other things, of the social responsibilities of scientists and of the many ramifications of information technology. Although information technology should be a means of bending technology to the service of human rights, it often tends to polarize the developing and developed worlds. Moreover, the computer is not merely a means of storage and dissemination of information but also an active tool in the generation of new technology.

Other areas of vital importance, because they obstruct the receipt of new technology, are the armaments trade, the politics of food, and the inadequacies of legal concepts and structures moulded to the needs and priorities of the developed world. These are briefly surveyed.

The chapter ends with some desiderata which will help in the achievement of the overall purpose of this project.

The problems of the environment are largely the result of technological "advance." They damage human rights at every level. If we are to turn technology to the affirmative science of human rights, one of the most vital areas to which we should be devoting our attention is the use of technology to repair or prevent environmental damage.

Dr Vukasovic addresses this problem in a chapter which probes the interdependence between the environmental issue and all human rights – the rights to life and health and all other social, economic, cultural, political, and civil rights. Not without reason is environmental law proving to be one of the most dynamic areas of both municipal and international law.

Environmental rights can be viewed both as collective and as individual rights. Thus, while for example the African Charter speaks of environmental rights as rights belonging to peoples, the Declaration of the Stockholm Conference on the Human Environment speaks of environmental rights also as individual rights. It is thus clearly a right belonging to both individuals and groups. This raises definitional problems as well as problems of collision of rights, and necessitates the creation of institutional machinery not only at the level of the UN but at regional and subregional levels.

Dr Vukasovic sees it as essential to the whole field of development of these newly emergent rights that they be more precisely defined and more efficiently regulated. This is an essential step in the promotion of human rights.

Apart from such conceptual and structural development there are numerous areas of great potential for technology to be harnessed in the service of environmental human rights. Some of these technologies exist in the military sphere and need to be used for developmental purposes. The use of space technology for remote sensing of the earth is an example; the technology for combating and preventing damage caused by ultra-hazardous activities is another.

Attention will need to be directed into areas of prevention and regulation, and this necessitates both the development of rights to information and duties of disclosure. People have the right to know the real state of the environment, and secrecy in matters critical for human existence needs be counteracted.

Here again, developing countries face a special problem from lack of personnel as well as of regulation structures. This calls for co-operation from the developed world. Technology assessment at national and international levels is vital but is not easy to achieve without such co-operation. Decision-making concerning technological policy must not be made in closed governmental circles. There must be voluntary and democratic involvement of people, improved international relations through human rights co-operation, a development of international humanitarian law, and a linkage of environmental law with concepts of sustainable development.

Within the broad contours mapped out in these chapters a number of potential research projects are contained. There is no doubt that the topic of this book offers one of the most vital areas for the development of human rights in this period when science and technology dominate nearly every aspect of our lives.

Seeing the importance of the topic and the multitude of areas of research lying within it, an informed choice is required regarding the next stage of this enterprise.

There is no doubt that it needs to proceed to the stage of more specific investigations of certain selected areas of immediate relevance to developing societies.

Such researches could be in the area of an exploration of the politico-economic decision-making process, which needs attention if self-reliance in technological decision-making is to be stimulated. It may necessitate a research project on existing structures of technological decision-making in developing societies. It may call for an investigation of developmental goals which those societies may pursue. It may also travel beyond individual societies into the areas of co-operation among developing societies, generating broader principles than studies of particular countries in isolation. There is an immense wealth of research topics awaiting attention. The task of informed decision in the midst of this embarrassment of riches is not an easy one.

But such researches are by themselves inadequate. We need also an investigation of the multitude of new structures, national and international, which will need to be devised or improved if the ideals considered in this book are to be implemented.

Here again the possibilities revealed by our study are multifarious. Structures that look back to the past so that we can learn from former mistakes, and structures which look forwards to the future, grappling with problems and technologies we can only dimly understand, lie at the extremes of a spectrum of options which necessitate a considered choice.

Especially within developing societies, where the need for surveillance or regulatory mechanisms is most urgent and acute, there is a dearth or a total absence of such mechanisms. Much thought needs to be given to this problem with a degree of special urgency, for as long as such structures are not in place every advantage will be taken of their absence to introduce technologies in the interests of private profit which impede the flow of human rights benefits to the bulk of the populations of developing nations.

We have dealt thus far with the broad socio-political setting and with the new institutional structures required.

However, the suggestions in this volume do not end there. We need also further revision and development, on the conceptual plane, of the human rights themselves. New formulations of rights and new refinements of existing rights will need consideration. Examples are the right of access to intellectually creative work and the right of participation in social decisions.

Moreover, new approaches will need to be made to the consideration of such concepts as the right to property (and intellectual property), the right to freedom of contract, and the right to academic freedom. There is work here on a theoretical plane for jurists and for human rights specialists.

But structures and concepts are of no value unless there are also personnel to make them work in directions of maximum value. Here we have the need to train personnel for all the purposes and procedures that may emerge from this study. Among them are the training of personnel to man the new structures, the carrying of the messages emerging from this study to the grass roots where they must be implemented, education in new attitudes of self-reliance and of activism to overcome traditional passivity and resignation to the power and inevitability of technology. New cadres of third-world personnel must emerge imbued with the resolve that technology can be harnessed in the service of their communities and their socio-economic goals.

We commend all these possibilities to the Human Rights Commission with the confidence that an informed choice will be made for the next research stage in this project. As we prepare to celebrate the fortieth anniversary of the Universal Declaration of Human Rights, there could be few projects worthier of continuing research than the study of the ways in which the most powerful force of this age can be turned to the service of its most powerful ideological declaration.

Appendix 1. Resolution 1986/9 of the Commission on Human Rights

1986/9. USE OF SCIENTIFIC AND TECHNOLOGICAL DEVELOPMENTS FOR THE PROMOTION AND PROTECTION OF HUMAN RIGHTS AND FUNDAMENTAL FREEDOMS

The Commission on Human Rights,

Recalling its resolutions 1983/41 of 9 March 1983 and 1984/27 of 12 March 1984,

Recalling once again the relevant provisions of the Proclamation of Tehran and the relevant resolutions of the General Assembly and the Commission on Human Rights concerning human rights and scientific and technological developments,

Recalling also the Vienna Programme of Action on Science and Technology for Development adopted by the United Nations Conference on Science and Technology for Development,

Convinced of the paramount importance of the application of science and technology to economic and social progress and to the promotion and enjoyment of human rights and fundamental freedoms,

Recognizing the need to extend the benefits of science and technological developments to the developing countries,

Noting that various useful studies have been undertaken by United Nations bodies in accordance with General Assembly resolution 2450 (XXIII) of 19 December 1968 and subsequent resolutions with respect to human rights issues arising from developments in science and technology,

Recognizing that the effects of scientific and technological developments on human rights and fundamental freedoms have both beneficial and harmful aspects and therefore must be examined in their totality,

Taking into account the reports of the Secretary-General prepared in accordance with Commission resolutions 1983/41 and 1984/27,

1. *Expresses its appreciation* to Member States and relevant international organizations which have submitted their views to the Secretary-General on the most effective ways and means of using the results of scientific and technological developments for the promotion and realization of human rights and fundamental freedoms,

2. *Calls upon* all States to make every effort to utilize the benefits of scientific and technological developments for the promotion and protection of human rights and fundamental freedoms,

3. *Invites* the United Nations University, in co-operation with other interested academic and research institutions, to study both the positive and the negative impacts of scientific and technological developments on human rights and fundamental freedoms and expresses the hope that the United Nations University will inform the Commission on Human Rights of the results of its study on the question.

Appendix 2. Declaration on the Use of Scientific and Technological Progress in the Interests of Peace and for the Benefit of Mankind

PROCLAIMED BY GENERAL ASSEMBLY RESOLUTION 3384(XXX) OF 10 NOVEMBER 1975

The General Assembly,

Noting that scientific and technological progress has become one of the most important factors in the development of human society,

Taking into consideration that, while scientific and technological developments provide ever increasing opportunities to better the conditions of life of peoples and nations, in a number of instances they can give rise to social problems, as well as threaten the human rights and fundamental freedoms of the individual,

Noting with concern that scientific and technological achievements can be used to intensify the arms race, suppress national liberation movements and deprive individuals and peoples of their human rights and fundamental freedoms,

Also noting with concern that scientific and technological achievements can entail dangers for the civil and political rights of the individual or of the group and for human dignity,

Noting the urgent need to make full use of scientific and technological developments for the welfare of man and to neutralize the present and possible future harmful consequences of certain scientific and technological achievements,

Recognizing that scientific and technological progress is of great importance in accelerating the social and economic development of developing countries,

Aware that the transfer of science and technology is one of the principal ways of accelerating the economic development of developing countries,

Reaffirming the right of peoples to self-determination and the need to respect human rights and freedoms and the dignity of the human person in the conditions of scientific and technological progress,

Desiring to promote the realization of the principles which form the basis of the Charter of the United Nations, the Universal Declaration of Human Rights, the International Covenants on Human Rights, the Declaration on the Granting of Independence to Colonial Countries and Peoples, the Declaration on Principles of International Law concerning Friendly Relations and Co-operation among States in accordance with the Charter of the United Nations, the Declaration on Social Progress and Development, and the Charter of Economic Rights and Duties of States;

Solemnly proclaims that:

1. All States shall promote international co-operation to ensure that the results of scientific and technological developments are used in the interests of strengthening international peace and security, freedom and independence, and also for the purpose of the economic and social development of peoples and the realization of human rights and freedoms in accordance with the Charter of the United Nations.
2. All States shall take appropriate measures to prevent the use of scientific and technological developments, particularly by the State organs, to limit or interfere with the enjoyment of the human rights and fundamental freedoms of the individual as enshrined in the Universal Declaration of Human Rights, the International Covenants on Human Rights and other relevant international instruments.
3. All States shall take measures to ensure that scientific and technological achievements satisfy the material and spiritual needs of all sectors of the population.
4. All States shall refrain from any acts involving the use of scientific and technological achievements for the purposes of violating the sovereignty and territorial integrity of other States, interfering in their internal affairs, waging aggressive wars, suppressing national liberation movements or pursuing a policy of racial discrimination. Such acts are not only a flagrant violation of the Charter of the United Nations and principles of international law, but constitute an inadmissible distortion of the purposes that should guide scientific and technological developments for the benefit of mankind.
5. All States shall co-operate in the establishment, strengthening and development of the scientific and technological capacity of developing countries with a view to accelerating the realization of the social and economic rights of the peoples of those countries.
6. All States shall take measures to extend the benefits of science and technology to all strata of the population and to protect them, both socially and materially, from possible harmful effects of the misuse of scientific and technological developments, including their misuse to infringe upon the rights of the individual or of the group, particularly with regard to respect for privacy and the protection of the human personality and its physical and intellectual integrity.
7. All States shall take the necessary measures, including legislative measures, to ensure that the utilization of scientific and technological achievements promotes the fullest realization of human rights and fundamental freedoms without any discrimination whatsoever on grounds of race, sex, language or religious beliefs.
8. All States shall take effective measures, including legislative measures, to prevent and preclude the utilization of scientific and technological achievements to the detriment of human rights and fundamental freedoms and the dignity of the human person.
9. All States shall, whenever necessary, take action to ensure compliance with legislation guaranteeing human rights and freedoms in the conditions of scientific and technological developments.

Contributors

Saneh Chamarik was born in 1927. He holds a Bachelor of Law degree from the University of Moral and Political Sciences, Bangkok, Thailand, and a Bachelor of Arts degree from Manchester University, United Kingdom. From 1950 to 1960 he was in the Political Department of the Thai Ministry of Foreign Affairs. From 1960 to 1987 he was a member of the Faculty of Political Science, Thammasat University. In 1975 and 1976 he was Vice-Rector of Thammasat University. Dr Chamarik was the President of the Social Science Association of Thailand from 1982 to 1985 and in the same years he was also the Director of the Thai Khadi Research Institute at Thammasat University. In 1985 Dr Chamarik was appointed Trustee of the Thailand Development Research Institute, which office he continues to hold. His publications include *Buddhism and Human Rights* (1979), *Some Thoughts on Human Rights Protection and Promotion*, *Thai Politics and Education* (1980, in Thai), and *Thai Politics and Constitutional Development* (1983, in Thai).

Tom J. Farer is Professor of Law and International Relations at the American University in Washington, D.C., and Professor of Law at the University of New Mexico, where he previously served as University President. He has taught at Columbia, Princeton, Rutgers, and the Johns Hopkins School of Advanced International Studies. He has been an adviser to the commander of the Police Force of the Somali Republic and a special assistant respectively to the General Counsel of the US Department of Defense and the US Assistant Secretary of State for Inter-American Affairs. He was for eight years a member of the Inter-American Commission on Human Rights of the OAS and served for two terms as Commission President. He has been a fellow of the Smithsonian Institution's Wilson Center. His books include *Warclouds on the Horn of Africa*, *The Grand Strategy of the United States in Latin America*, and *US Ends and Means in Central America*.

Amilcar O. Herrera is Director of the Institute of Geosciences and Centre of Science and Technology Policy of the Universidade Estadual de Campinas, Brazil. Currently, he is also the Co-ordinator of the Technological Prospective for Latin America Project, sponsored by the UNU and IDRC. Dr Herrera has been associated with the University of Buenos Aires, the University of Chile, and the University of Sussex (SPRU), as well as the Fundacion Bariloche in Argentina and the United Nations University. He worked in the field of Economic Geology and National Resources at the University of Buenos Aires until 1969, and later worked in the field of Science and Technology Policy. In addition, he has been the Director of the Latin American World Model, made as an answer to the MIT model "The Limits of Growth." Dr Herrera has written extensively and has published books and articles on economic geology and science and technology policy.

Yo Kubota was born in 1951. Human rights research suffered a serious loss with his untimely death while on UN duty in Namibia. He was formerly a Lecturer in Law at Nihon University and held the position of Human Rights Officer under the United Nations Organization from January 1981. He served as Secretary to the United Nations Sub-Commission on the Prevention of Discrimination and Protection of Minorities, the UN Working Group on Contemporary Forms of Slavery, and other bodies in the United Nations Centre for Human Rights. He was also a visiting Professor at Meiji University School of Law and the Hosei University School of Law in Tokyo. He was the Rapporteur to the Steering Committee of the United Nations University on Science and Technology and Human Rights. Dr Kubota was the author of several books and a number of articles in the fields of human rights and international law.

Shigeru Nakayama was born in Japan in 1928 and graduated from the Department of Astronomy, University of Tokyo in 1951. He worked with Thomas Kuhn at Harvard University and Joseph Needham at Cambridge University. He was awarded the degree of Doctor of Philosophy by Harvard University in 1959 for his work on the history of science and learning. Dr Nakayama is currently a Professor of the College of Arts and Sciences at the University of Tokyo. His published work includes work on the development of knowledge and the future of research.

Vid B. Vukasovic is Research Fellow at the Institute of International Politics and Economics in Belgrade, Yugoslavia. From 1984 to 1987 he was Head of the International Law Department. He has taken part in a number of research projects, conferences, etc., on both national and international levels, either in his private capacity or as a governmental expert or representative of professional associations, among them the International Law Association and the Yugoslav Law Association. He has lectured widely and been a member of various boards of editors. Dr Vukasovic's books include *International Law and the Development of Science and Technology* (Belgrade, 1978), *Protection and Improvement of the Environment* (Belgrade, 1980), and *Environmental Protection and the United Nations* (Belgrade, 1985). Dr Vukasovic has published many articles and other shorter works dealing with environmental protection, development of science and technology, human rights, the work of international organizations, and related subjects.

C.G. Weeramantry was a Judge of the Supreme Court of Sri Lanka from 1967 to 1972, and since 1972 has been Professor of Law at Monash University, Melbourne. He is currently the Chairman of the Commission of Inquiry into the Rehabilitation of the Worked-out Phosphate Lands in Nauru. He is a Doctor of Laws of the University of London, and an Associate Academician of the International Academy of Comparative Law. His published works include *The Law of Contracts, being a Comparative Study of Roman-Dutch, English and Customary Law relating to Contracts* (2 vols.); *The Law in Crisis*; *Equality and Freedom: Some Third World Perspectives*; *Apartheid: The Closing Phases?*; *The Slumbering Sentinels: Law and Human Rights in the Wake of Technology*; *Human Rights in Japan*; *Law: The Threatened*

Peripheries; An Invitation to the Law; Nuclear Weapons and Scientific Responsibility; and *Islamic Jurisprudence – Some International Perspectives.*

Hiroko Yamane studied Political Science and Law at Tokyo University as well as Yale University, the University of Paris, and other institutions of higher learning. She has been on the staff of the Division of Human Rights and Peace at Unesco since 1978. Dr Yamane has been Associate Professor of International Law and International Relations at Meiji Gakuin University, Tokyo, since 1987.

vencendo

gigantes

HERNANDES DIAS LOPES

vencendo gigantes

HAGNOS

Revisão
Heloisa Wey Neves Lima

Capa
Alexandre Gustavo

Diagramação
Atis design ltda

1ª edição - fevereiro de 2005
Reimpressão - julho de 2005
Reimpressão - junho de 2007
Reimpressão - julho de 2008

Coordenador de produção
Mauro W. Terrengui

Impressão e acabamento
Imprensa da Fé

Todos os direitos reservados para:
Editora Hagnos
Av. Jacinto Júlio, 27
04815-160 - São Paulo - SP - Tel (11)5668-5668
hagnos@hagnos.com.br - www.hagnos.com.br

Dados Internacionais de Catalogação na Publicação (CIP)
(Câmara Brasileira do Livro, SP, Brasil)

Lopes, Hernandes dias -
Vencendo gigantes / Hernandes Dias Lopes;
São Paulo, SP: Hagnos 2005

ISBN 85-89320-61-8

1. Conduta de vida 2. Encorajamento 3. Motivação 4. Perseverança (Ética)
5. Sucesso 6. Vida cristã I. Título

04-7777   CDD-248.4

Índices para catálogo sistemático:
1. Perseverança: Vida cristã 248.4
2. Sucesso: Vida cristã 248.4
3. Vitória: Vida cristã 248.4

Índice

Dedicatória

Dedico este livro ao Rev. Ronaldo de Almeida Lidório, missionário preparado por Deus para grandes batalhas, vencedor de gigantes, exemplo para os fiéis, amigo, conselheiro e consolador, homem segundo o coração de Deus.

Introdução

Vencedores de gigantes não nascem prontos, eles se tornam assim. Nesse processo não existe sorte nem azar, mas coragem, determinação e muito trabalho. Thomas Alva Edison costumava dizer que nossas conquistas são resultado de 10% de inspiração e 90% de transpiração. Vencedores de gigantes não olham para os obstáculos, mas para as oportunidades. Henry Ford estava coberto de razão quando afirmou: "Obstáculos são aquelas coisas tenebrosas que vemos quando desviamos os olhos do nosso objetivo".

Vencedores de gigantes triunfam em público, porque já tiveram importantes vitórias privadas. Saul destacava-se do ombro para cima de todos os demais guerreiros de Israel, mas não tinha estatura suficiente para lutar contra o gigante Golias. Davi enfrentou e venceu o gigante porque já triunfara sobre um leão e um urso na solidão das montanhas de Belém. O que você é quando está sozinho determina o que você será publicamente. O caráter precede a performance. Ser é mais importante do que fazer.

Vencedores de gigantes não são covardes. Eles não fogem e nunca desistem. São perseverantes. Eles não se contentam com nada menos que a vitória. Fazem das derrotas do passado experiências indispensáveis para construírem as vitórias do futuro. Experiência não é simplesmente aquilo que acontece na vida de uma pessoa, mas como ela lida com o que lhe acontece. Os vencedores nunca desviam os olhos do alvo. Eles são determinados e obcecados pela vitória, não importa o quanto possa lhes parecer difícil alcançá-la. Quando Thomas Alva Edison inventou a lâmpada elétrica, foram necessárias mais de duas mil experiências antes de obter sucesso. Certa vez, um jovem repórter perguntou como ele se sentia tendo fracassado tantas vezes. Ele respondeu: "Eu nunca fracassei. Inventei a lâmpada elétrica. Só que para chegar lá, foi preciso uma caminhada de dois mil passos". John Milton ficou cego aos quarenta e quatro anos de idade. Dezesseis anos depois, escreveu o clássico *Paraíso Perdido*. Após uma perda progressiva da audição, o compositor alemão Ludwig Van Beethoven ficou totalmente surdo aos quarenta e seis anos. Apesar disso, continuou compondo, e algumas de suas melhores composições, inclusive cinco sinfonias, foram escritas durante seus últimos anos. Quando Alexander Graham Bell inventou o telefone, em 1876, não causou grande impressão entre os possíveis patrocinadores. Depois de usar o aparelho para dar um telefonema apenas como demonstração, o Presidente Rutherford Hayes disse: "É uma invenção surpreendente, mas quem se interessaria em usá-lo?"

Em 1952, Edmund Hillary tentou escalar o Monte Everest, a mais alta montanha do mundo, com 8.800 metros. Algumas semanas depois de uma tentativa fracassada, pediram-lhe que falasse para um grupo de pessoas na Inglaterra. Propositadamente, colocaram uma grande gravura do Monte Everest na parede do auditório. Quando o alpinista viu a gravura, abaixou a cabeça e caminhou para o palco, cabisbaixo. Mas, ao lhe perguntarem se havia desistido, ele se levantou, apontou para a figura da montanha no fundo do palco e disse em voz alta: "Monte Everest, dessa vez você me derrotou, mas na próxima eu o derrotarei, porque você já cresceu tudo o que tinha de crescer, mas eu ainda estou crescendo!" Em maio do ano seguinte, Edmund Hillary tornou-se o primeiro homem a escalar o Monte Everest. Vencedores de gigantes não desistem, mesmo diante dos maiores obstáculos.

Vencedores de gigantes não são movidos por elogios nem desencorajados pelas críticas. Eles não são hipersensíveis ou melindrosos. São livres para escolher seu próprio caminho. Vicktor Frankl dá o seu testemunho: "Todos que, como eu, passaram pelos campos de concentração, podem se lembrar daqueles que iam de tenda em tenda, confortando as pessoas, dando-lhes seu último pedaço de pão. Talvez fossem poucos, mas são prova suficiente de que se pode tirar tudo de um homem, menos uma coisa – a liberdade de escolher seu comportamento, quaisquer que sejam as circunstâncias, e de escolher seu próprio caminho". Você é livre para tomar suas decisões. É você quem escolhe se quer ou não vencer as crises e derrotar os gigantes. Mesmo que o reconhecimento não venha e as críticas se multipliquem, não abra mão de ser um vencedor. Enfrente essa peleja confiante na vitória.

Convido você a caminhar comigo nessa aventura. Você pode ser um vencedor de gigantes. Não fuja da luta, não se esconda, enfrente-a e vença os gigantes!

1

Identificando os gigantes que nos assustam

Os gigantes têm a capacidade de nos assustar.

Os gigantes existem, eles são reais, e estão espalhados por toda parte. Eles sempre cruzam o nosso caminho para nos desafiar e instilar medo em nossos corações. São numerosos, opulentos, atrevidos e escarnecedores. Sempre que aparecem, tripudiam sobre nós com grande insolência.

Ninguém é capaz de obter grandes vitórias sem que antes, em algum lugar, em algum tempo, em diversas e adversas circunstâncias, tivesse que enfrentar terríveis gigantes. Esses gigantes podem ser pessoas, hábitos, sentimentos, circunstâncias e críticas. É quase impossível identificar todos eles.

Há gigantes reais e gigantes fictícios. Uns nos atacam por fora, outros por dentro. Muitos deles existem apenas em nossa imaginação. Nós os criamos, e eles acabam se tornando mais fortes que nós. Fabricamos esses monstros no laboratório do medo e eles se levantam contra nós para nos atormentar. Muitas vezes, tememos mais os gigantes invisíveis do que aqueles que nossos olhos alcançam. Eles tomam o castelo da nossa mente, se assentam no trono das nossas emoções e nos castigam impiedosamente.

Passei minha infância numa região rural. Desde a mais tenra idade ouvi contar histórias pavorosas de monstros, casas mal-assombradas e pessoas mortas que apareciam para aterrorizar os vivos. Quando uma pessoa conhecida morria, eu ficava um bom tempo sem sair de casa. Tinha pavor de passar na frente do cemitério. Na verdade, o medo era tanto que quando eu caminhava no escuro, as mais estranhas figuras se desenhavam diante dos meus olhos. Todos esses medos foram crescendo dentro de mim e se transformando em gigantes que custei a vencer. Esses gigantes existiam apenas na minha imaginação, mas a aflição que eu sentia era real.

O medo e a ansiedade assolam crianças e adultos, doutores e analfabetos, homens do campo e da cidade, indiscriminadamente. As pessoas se tornam prisioneiras das fobias que elas mesmas criam. Temem a vida e a morte, o casamento e o divórcio. Temem a solidão, e temem compartilhar a vida com outra pessoa. Temem a multidão e a privacidade, o trabalho e o desemprego, o presente e o futuro. Temem o tempo e a eternidade. Enquanto esses gigantes se tornam cada vez maiores, nós nos encolhemos, esmagados, debaixo de suas botas.

A fobia é um gigante real. Alguns têm medo de lugares fechados, outros de lugares abertos. Uns sofrem de claustrofobia, outros de agorafobia. Percebi que a claustrofobia era um gigante em minha vida quando visitei o Egito pela primeira vez, em 1986. Eu e alguns companheiros de viagem resolvemos conhecer o túmulo do faraó, no interior da pirâmide de Quéops. Consideradas uma das sete maravilhas do mundo antigo, as pirâmides sempre despertaram a admiração dos viajantes, desconhecidos ou famosos, que chegavam às margens do rio Nilo, especialmente as três mais importantes, Quéops, Quéfren e Miquerinos. Essas construções monumentais eram usadas como túmulos para os faraós do Egito.

Aventurei-me a subir por aqueles corredores escuros, entupidos de gente curiosa, enfrentando o calor e o ar viciado para chegar à famosa sala onde se encontrava o túmulo do faraó. Ao longo da escalada fiquei inquieto ao ver algumas pessoas aflitas, com a respiração ofegante, desistirem da peregrinação. De repente, atordoado com a situação, dei-me conta de que o ar fugira dos meus pulmões. Entrei em pânico. Sabia que naquele momento seria impossível qualquer atendimento de emergência ou uma fuga imediata do local. Aflito, pensei: "É o fim. Vou morrer asfixiado nas entranhas desse monumento de pedras milenares". Clamei por socorro ao Ronaldo Gama, um amigo que me acompanhava, e ele conseguiu me tranqüilizar. Fechei os olhos, procurando desligar-me mentalmente daquele lugar abafado e completamente fechado em que me encontrava. Aos poucos, o fôlego da vida começou novamente a pulsar em meu peito e eu ganhei alento para prosseguir. Descobri naquele dia que lugares hermeticamente fechados são verdadeiros gigantes para mim.

O medo é mais que um sentimento, é um espírito (2 Tm 1.7). Esse espírito oprime, escraviza e destrói suas vítimas. O medo é um gigante que desafia a nossa alma. Somos assaltados muitas vezes por perigos reais, mas as situações irreais também nos assustam. Certo homem tinha muito medo de passar defronte de cemitérios. Nem mesmo a idade adulta havia conseguido libertá-lo dos fantasmas que povoavam sua mente. Um dia, sua empresa o transferiu para uma distante cidade. O único problema é que a casa onde deveria morar ficava atrás de um cemitério. O homem ficou desapontado com a notícia, mas pensou que poderia ser sua grande chance de vencer aquele gigante invisível. Durante algum tempo, sempre que aquele homem passava pelo cemitério sozinho, à noite, ele tremia de medo. Contudo, com o tempo, o medo foi se dissipando e dentro de alguns meses, aquele gigante já fazia parte do seu passado. A vitória sobre o medo foi tal que ele começou a cortar caminho para sua casa passando por dentro do cemitério. Numa noite particularmente escura, ao passar pela trilha dentro do cemitério, ele caiu numa cova recém-aberta. Tentou em vão sair daquele buraco, até que por volta de meia-noite, pensou: "Não adianta gritar por socorro aqui no cemitério a esta hora da noite. É melhor esperar o dia amanhecer aqui mesmo". Mas ele não esperava que outra pessoa passasse por aquele mesmo caminho e caísse dentro da cova também. Depois de ter tentado inutilmente sair daquela sepultura, a segunda pessoa sentou-se, exausta, quando de repente, ouviu uma voz que vinha da escuridão lhe dizer: "É companheiro, dessa você não sai, não!" Impulsionado pelo medo, o homem deu um salto descomunal, além do que suas forças lhe permitiam, e conseguiu sair da cova. O medo fez com que ele superasse seus próprios limites, despejando uma dose extra de adrenalina no sangue e permitindo-lhe alcançar uma façanha inusitada. De fato, o medo provocado por situações irreais nos esmaga e nos aflige tanto ou mais do que aquele que resulta de fatos reais.

O medo é um sentimento democrático, que atinge a todos, crianças e velhos, doutores e analfabetos, cosmopolitas e campesinos. As pessoas se tornam prisioneiras das fobias que elas mesmas criam e desenvolvem. Outras se tornam cativas de monstros fictícios que foram emoldurados no seu imaginário mais recôndito. Não são poucas as pessoas que sofrem não por motivos evidentes, mas por motivos que não conhecem, que apenas ouviram falar. Essas pessoas fantasiam tal situação de pavor que essa condição fictícia torna-se realidade diante dos seus olhos e elas passam a sofrer como se estivessem no fragor de uma situação realmente desesperadora.

A ansiedade é outro gigante que mantém as pessoas em cativeiro. Você conhece alguma pessoa ansiosa? Tenho certeza que sim, afinal você já deve ter se olhado no espelho hoje. A ansiedade é o estrangulamento da alma, o flagelo das emoções. Ansiedade significa estrangular, sufocar. Traz também a idéia de rasgar, provocar uma fissura interior, abrir um abismo na alma, cavar um vale nas emoções, rasgando-nos ao meio e produzindo em nós uma esquizofrenia existencial.

Não apenas o medo, mas também a ansiedade é outro gigante que apavora muitas pessoas. A ansiedade é um gigante que intimida grandes e pequenos, citadinos e rurícolas, reis e vassalos, ricos e pobres, homens e mulheres. Certa feita, abordei uma criança, parabenizando-a pela placidez da sua vida. Disse-lhe: "Que vida boa, hein?!". Ela me respondeu: "O senhor é que pensa!" Até as crianças abrigam no coração a ansiedade.

A ansiedade é uma espécie de estrangulamento da alma. Ela nos sufoca, nos tira o fôlego. A ansiedade é um sentimento errado. Ela ataca tanto a razão quanto as emoções. Ela mexe com a nossa cabeça e com o nosso coração. Ela nos leva à lona e pisa em nosso pescoço.

A ansiedade é inútil. Ela não pode alterar a situação. Não podemos acrescentar nem um côvado à nossa vida, por maior que seja a nossa ansiedade. Além de não mudar as circunstâncias, ela nos enfraquece, roubando-nos o vigor. Em vez de nos preparar para enfrentar os problemas, a ansiedade nos aprisiona ao passado, esvazia o nosso presente e nos impede de vencer no futuro.

A ansiedade é tolice. A pessoa ansiosa antecipa o sofrimento e curte a dor antes do problema chegar. Não raro, a pessoa ansiosa sofre desnecessariamente por um problema que nem chega a acontecer.

A ansiedade revela incredulidade. Onde há ansiedade não há lugar para a fé. O gigante da ansiedade só oprime aqueles que duvidam que Deus está no controle. Ansiedade é tirar os olhos de Deus para colocá-los nas circunstâncias. Os dez espias de Israel foram esmagados pela síndrome do gafanhoto porque tiraram os olhos de Deus para colocá-los na fortaleza dos gigantes. Isso lhes deu uma perspectiva falsa do problema, fazendo com que dissessem: os gigantes são fortes e nós fracos; são muitos e nós poucos; são guerreiros e nós nômades; são gigantes e nós gafanhotos. Porque duvidaram da promessa de Deus, sentiram-se menos do que príncipes, menos do que homens, sentiram-se insetos, gafanhotos. Eles não só introjetaram um sentimento de fracasso na alma, mas também contaminaram todo o arraial de Israel com o seu pessimismo. Como resultado, toda aquela geração pereceu no deserto e não pôde entrar na Terra Prometida. Apenas Josué e Calebe olharam para as circunstâncias com os olhos da fé. Note o que eles disseram: *"Certamente o Senhor nos deu toda esta terra nas nossas mãos, e todos os seus moradores estão desmaiados diante de nós"* (Js 2.24). Eles creram e venceram. Por terem crido, eles tomaram posse da Terra Prometida.

Não olhe ao redor, para o vento que ruge ou para o bramido das ondas. Olhe para o Senhor. Pedro andou sobre as águas enquanto manteve os olhos fitos em Jesus. O milagre estava firme debaixo dos seus pés. Mas, num instante, a fúria do vento, o barulho das ondas encheram o seu coração de medo e ele tirou os olhos de Jesus e afundou.

Há muitos outros gigantes que ainda nos afligem.

Enfrentamos o gigante da crise financeira. O mercado de trabalho está cada vez mais exigente e cada vez mais escasso. Nessa aldeia global os ricos tornam-se cada vez mais opulentos e os pobres cada vez mais desesperados. O mercado global está cada vez mais guloso, exigindo mais do nosso dinheiro e do nosso tempo. Consumimos hoje cinco vezes mais do que a geração de 1950. Os luxos de ontem são as necessidades de hoje. Hoje, mais de 70% das famílias precisam dispor de duas rendas para manter o mesmo padrão de vida de vinte e cinco anos atrás, quando mais de 70% das famílias viviam com apenas uma renda. As coisas estão tomando o lugar das pessoas. Os bens materiais estão tomando o lugar dos relacionamentos. Sacrifica-se no altar da realização profissional o casamento e a própria vida dos filhos. Cerca de 50% dos casamentos terminam em divórcio. Dez anos depois do segundo casamento, 70% buscam o divórcio novamente. Temos presenciado casais amancebados, descasados e divorciados atolando-se cada vez mais na crise econômica, tendo que repartir o salário entre duas famílias.

A crise financeira alarga a base da pirâmide, enquanto alimenta o sistema guloso do consumismo. As riquezas concentram-se nas mãos de poucos. Nesse mundo de economia global, os grandes engolem os pequenos. Cerca de 50% das riquezas do planeta se acumulam nas mãos de algumas centenas de megaempresas. Há empresas mais ricas do que muitos países. A *General Motors* é mais rica que a Dinamarca. A *Toyota* é mais rica do que a África do Sul. A *Ford* é mais rica do que a Noruega. A *Wal-Mart* é mais rica do que cento e sessenta e um países. Quando John Rockefeller tornou-se o primeiro bilionário do mundo, isso foi noticiado como algo fantástico. Hoje há mais de quatrocentos bilionários só nos Estados Unidos. Bill Gates tem uma renda líquida de quatrocentos milhões de dólares por semana. Ele controla hoje 90% do mercado de computadores do mundo. Como encontrar um lugar ao sol neste mundo dominado pelos poderosos? A luta para conseguir um emprego é atualmente mais apertada do que para obter uma vaga na faculdade. Muitos jovens entram na univerdade cheios de sonhos e saem com o coração cheio de ansiedade.

A crise, porém, é uma encruzilhada que exalta uns e abate outros. Enquanto uns colocam os pés na estrada da vitória, outros descem a ladeira do fracasso. Enquanto uns se contentam com a visão do retrovisor, enxergando apenas o que ficou para trás, outros têm a visão aumentada pela luz do farol alto, enxergando largos horizontes por sobre os ombros dos gigantes. A crise revela os medíocres, mas descobre e exalta os heróis. É do ventre da crise que nascem os verdadeiros vencedores.

Para vencer a crise é preciso ter criatividade e coragem. John Rockefeller disse que o futuro não é fruto do acaso, ele é criado por homens de visão. Os visionários fazem da crise um tempo de oportunidade, enquanto os medíocres ouvem os embaixadores do caos, os profetas do pessimismo, e se alimentam da crise. Os vencedores agarram a crise pelo pescoço e fazem dela um campo de semeadura. Os vencedores semeiam nos desertos, transformando-os em pomares frutuosos. Os medrosos vêem os gigantes e correm para se esconder; os que têm pressa para vencer correm na direção deles para derrotá-los. As portas estão fechadas para os que enxergam a vida pela lente dos medrosos e pessimistas. Mas aqueles que sonham os sonhos de Deus transformam a crise em combustível motivacional para alargar seus horizontes.

Não há idade para sonhar e tentar novos planos e projetos. Winston Churchill assumiu o cargo de primeiro-ministro da Inglaterra aos sessenta e seis anos de idade, liderando a frente vitoriosa que barrou o avanço sanguinário de Adolf Hitler durante a Segunda Guerra Mundial. Roberto Marinho, presidente das Organizações Globo, estava com sessenta e dois anos de idade quando começou um novo projeto em sua vida, tornando-se um dos maiores empresários de comunicação do mundo. A Bíblia fala que Abraão começou a andar com Deus aos setenta e cinco anos de idade e estava com cem anos quando recebeu em seus braços o filho da promessa. Quando a maior parte das pessoas já "pendurou as chuteiras" há muito tempo, Abraão estava apenas começando uma gloriosa aventura com Deus. Moisés estava com oitenta anos de idade quando foi chamado por Deus para a maior missão da sua vida, libertar o povo de Israel do cativeiro do Egito. Calebe, aos oitenta e cinco anos de idade, estava ainda cheio de vigor e sonhos, entusiasmado com os planos para tomar a cidade de Hebrom e conquistar novos horizontes em sua vida. Na verdade, os maiores desafios de nossas vidas não ficaram para trás, mas estão à nossa frente. Ainda temos gigantes para derrubar e vitórias para celebrar.

2

A insolência dos gigantes que nos desafiam

Então saiu do arraial dos filisteus um homem guerreiro, cujo nome era Golias, de Gate, da altura de seis côvados e um palmo. Trazia na cabeça um capacete de bronze e vestia uma couraça de escamas cujo peso era de cinco mil siclos de bronze. Trazia caneleiras de bronze nas pernas e um dardo de bronze entre os ombros. A haste da sua lança era como o eixo do tecelão, e a ponta da sua lança, de seiscentos siclos de ferro; e diante dele ia o escudeiro [...] Disse mais o filisteu: Hoje afronto as tropas de Israel. Dai-me um homem, para que ambos pelejemos [...] Chegava-se, pois, o filisteu pela manhã e à tarde; e apresentou-se por quarenta dias (1 Sm 17.4-7,10,16).

A Bíblia fala de uma batalha que foi travada entre israelitas e filisteus. Os exércitos estavam acampados em dois montes, separados por um vale. Do acampamento dos filisteus saiu Golias, um gigante duelista, com mais de três metros de altura, usando uma armadura que pesava mais de oitenta quilos e uma lança cuja ponta pesava mais de doze quilos. Esse guerreiro amedrontador desafiou os soldados israelitas durante quarenta dias, afrontando-os e humilhando-os insolentemente, indagando de manhã e à tarde, se não havia pelo menos um homem para lutar com ele. As tropas de Israel ensarilhavam as armas e batiam em retirada, tomadas pelo medo.

Podemos extrair algumas lições desse episódio.

A primeira coisa que observamos aqui é que *diante da ameaça dos gigantes, o ânimo do povo se abate*. Os soldados de Saul ficaram aterrados de medo. Eles foram derrotados pelo medo antes mesmo de fugirem de Golias. O coração deles desmaiou, e eles se encolheram, acovardados. Tudo que conseguiam ver pela frente era o vexame e a vergonha por fugirem continuamente do adversário cheio de bravatas.

Sempre que supervalorizamos o poder dos gigantes que nos ameaçam, nós acabamos fugindo. Talvez você esteja caminhando por essa estrada da fuga há muito tempo. Você acorda de manhã, veste a farda, reúne seus exércitos, toca a trombeta, empunha as armas e enfileira-se para a batalha, mas assim que você ouve a voz do gigante, começa a tremer dos pés à cabeça e foge desesperado.

Golias desafiou os soldados de Saul oitenta vezes. Durante quarenta dias, pela manhã e à tarde, ele lançava o desafio. O moral dos soldados já estava no chão. Eles já estavam desacreditados até mesmo aos seus próprios olhos. Para eles, a vitória era um sonho ilusório, uma possibilidade absolutamente remota. Talvez você também já tenha desistido de acreditar na vitória. Você acorda, cumpre seu ritual diário, mas na hora da peleja você foge, em vez de enfrentar o seu gigante. Você já se acostumou a fugir. Suas armas estão ficando enferrujadas. Você as carrega, mas não as usa. E quando o dia termina, a voz desafiadora do gigante ainda fica retinindo em seus ouvidos.

A segunda coisa que notamos é que *os gigantes não apenas parecem ser inimigos imbatíveis, mas também são insolentes*. Os gigantes nos afrontam e zombam da nossa força, da nossa fé, do nosso Deus.

Eles escarnecem da nossa fraqueza, tripudiam sobre nossas armas e insultam o nosso Deus. Eles não têm nenhum respeito pelas nossas convicções, escarnecem e zombam da nossa religião. Eles não querem apenas nos humilhar, mas também banir Deus da nossa mente. Quando fugimos, os gigantes entendem que estão prevalecendo não apenas contra nós, mas também contra o nosso Deus.

A terceira é que *os gigantes parecem inatingíveis*. Golias trajava uma armadura cheia de escamas, impenetrável ao fio da espada. Além da armadura, ele usava caneleiras e capacete, e trazia um dardo no ombro e uma lança na mão. Como se não bastasse, um escudeiro sempre ia à sua frente para protegê-lo. Os gigantes se preparam para a luta, se protegem e se escondem atrás de estruturas e esquemas impenetráveis. São humanamente inatingíveis. Eles têm escudos de bronze, e se acham imunes aos ataques. Eles se abrigam debaixo da proteção da lei e se escondem sob o manto dos poderosos. Eles fazem as leis, e depois as manipulam, procurando brechas para poder escapar. De fato, eles são a lei.

Os gigantes se escondem atrás dos esquemas de corrupção que se instalam nas instituições e nos poderes constituídos. Muitas vezes, eles usam togas e assentam-se nos tribunais. Em vez de defenderem o povo e preservarem os valores morais que sustentam a nação, tornam-se ratazanas famintas, sanguessugas insaciáveis e parasitas que se alimentam do sangue do próprio povo. Em vez de abençoar o povo, eles o devoram, como bestas-feras. Eles oprimem, mentem, corrompem, matam e escapam. Esses gigantes não caem pelas armas convencionais. Seus escudeiros os livram dos ataques. As armaduras que vestem são impenetráveis. Por isso são tão arrogantes e insolentes. Sentem-se inexpugnáveis e invencíveis.

A quarta lição nos diz que *os gigantes são persistentes*. Golias desafiou os exércitos de Saul durante quarenta dias. Os soldados israelitas ficaram com o ânimo abatido depois de tantos fracassos. Eles se acostumaram a fugir, e tornaram-se o próprio retrato do fracasso. Se os gigantes que estão à sua frente conseguem desviar os seus olhos

Vencendo Gigantes

de Deus, se a crise faz com que você olhe só para suas limitações e fraquezas, então sua derrota, de fato, já está lavrada. Por fim, *os gigantes precisam ser enfrentados e vencidos*. Não podemos buscar um atalho e fugir deles, porque estão por todos os lados e em cada esquina. Os gigantes pensam que são imbatíveis, por isso não saem do nosso caminho. Eles não se intimidam com o nosso aparato. Os gigantes, na verdade, revelam quem realmente somos: covardes ou corajosos. Os espias de Israel se dividiram em dois grupos. Ambos viram as mesmas circunstâncias, estavam no mesmo projeto, mas dez disseram que os gigantes eram invencíveis e apenas dois disseram que eles, e não os gigantes, eram imbatíveis. Um grupo viu no poder dos gigantes a sua fraqueza; o outro, olhou para cima, para Deus, e viu um vencedor de gigantes. A crise é assim: revela uns e abate outros; promove uns e derrota outros; desmascara os covardes e aponta os heróis. É do útero da crise que despontam os grandes vencedores. Eles não se intimidam com o tamanho dos gigantes, são obcecados pela vitória e não descansam enquanto não vêem os gigantes beijando o pó.

Os vencedores são visionários, enxergam mais longe, vêem o invisível, crêem no impossível e tocam o intangível. Eles não olham a vida pelas lentes embaçadas do pessimismo, mas vêem a vida pelo telescópio das ilimitadas possibilidades de Deus. A diferença entre um vencedor e um perdedor é a visão, a perspectiva e não as circunstâncias. A história a seguir ilustra bem esse ponto. Um homem de vida desregrada, ébrio, perdulário, teve dois filhos. Um deles tornou-se um grande empresário, respeitado por todos, homem de bem e de caráter impoluto. O outro, enveredou pelos caminhos tortuosos dos vícios, destruindo sua vida de tal forma que foi parar na prisão.

Um repórter, intrigado com o fato desses dois irmãos terem destinos tão diferentes, apesar de terem sido criados no mesmo lar, sob a influência do mesmo pai e recebendo a mesma herança genética, resolveu entrevistá-los. O primeiro a ser entrevistado foi

24

A insolência dos gigantes que nos desafiam

o que estava na prisão. O repórter perguntou então àquele homem sucateado pelo vício:

> – Como você veio parar na prisão? Por que você se entregou aos vícios de forma tão degradante?

Ele respondeu:

> – Com o pai que eu tive, poderia ser diferente?

O repórter saiu dali e foi procurar o outro filho, fazendo-lhe a seguinte pergunta:

> – Como você chegou onde está? Qual o segredo do seu sucesso?

E ele respondeu:

> – Com o pai que eu tive, não poderia seguir seus passos. Assim, dediquei-me com todo ardor a um ideal elevado e me tornei um vencedor.

A visão determina a ação. Os verdadeiros gigantes não estão fora de nós, mas dentro de nós. Não somos derrotados pelas circunstâncias, mas pelos nossos sentimentos. Os gigantes fictícios e irreais são mais fortes que os gigantes reais. Eles nos atacam por fora, mas os sentimentos por dentro. Como podemos vencer esses gigantes? A história de Davi e Golias lança luz sobre esse magno assunto. Veremos a resposta a essa questão nos próximos capítulos.

3

Vencedores de gigantes não ouvem a voz dos pessimistas

Então saiu do arraial dos filisteus um homem guerreiro, cujo nome era Golias, de Gate, da altura de seis côvados e um palmo [...] Clamou às tropas de Israel e disse-lhes: Para que saís, formando-vos em linha de batalha? Não sou eu filisteu e vós servos de Saul? Escolhei dentre vós um homem que desça contra mim [...] Hoje afronto as tropas de Israel. Dai-me um homem, para que ambos pelejemos. Ouvindo Saul e todo o Israel estas palavras do filisteu, espantaram-se e temeram muito [...] Todos os israelitas, vendo aquele homem, fugiam de diante dele, e temiam grandemente (1 Sm 17. 4,8,10,11,24).

Alexandre Rangel, em seu livro *As Mais Belas Parábolas de Todos os Tempos,* conta a história de um homem que vivia à beira da estrada e vendia cachorros-quentes. Não tinha rádio e, por deficiência de visão, não podia ler jornais. Em compensação, vendia bons cachorros-quentes. A maneira como divulgava seu produto era bem simples: um cartaz na beira da estrada, anunciando a

mercadoria, e quando alguém passava, ele gritava: "Olha o ca-
chorro-quente especial!". E as pessoas compravam. Com isso, suas
vendas foram crescendo e os pedidos de pão e salsicha aumentando.
Entusiasmado, ele resolveu construir uma mercearia e telefonou
para o filho, que morava em outra cidade, para contar a novidade.
Mas, ao ouvir os planos do pai, o filho retrucou:

— Pai, o senhor não tem ouvido as notícias no rádio? Não tem
lido os jornais? O país está passando por uma crise muito séria, e a
situação internacional está muito instável!

Diante disso, o pai pensou: "Meu filho estuda na universidade,
ouve rádio e lê jornais, portanto, deve saber o que está dizendo!"
Assim, ele reduziu os pedidos de pão e salsicha, tirou o cartaz
da beira da estrada e não ficou mais apregoando os seus cachorros-quentes. Com isso, as vendas caíram do dia para a noite. O
pai, então, disse ao filho: "Você tinha razão, meu filho, a crise é
realmente muito séria!"

Cuidado com o vozerio dos pessimistas, eles olham para a
vida com lentes embaçadas. Eles perdem a vida por medo de
viver. Eles correm até da sombra. Eles pensam que os gigantes são
invencíveis.

O gigante Golias julgava-se imbatível. Era um guerreiro
inveterado, assombroso, experiente, que infundia medo em todo o
exército de Saul. Ele desafiou os exércitos de Israel e fez Saul e seus
soldados recuarem, empapuçados de medo. Durante quarenta dias,
Golias afrontou os exércitos do Deus vivo e não apareceu ninguém
com coragem para enfrentá-lo.

Mas, em meio à orquestra do medo, ouve-se o clarinete da
esperança. Entre o povo circulou a notícia de que o jovem pastor
Davi, que acabara de chegar de Belém, estava disposto a enfrentar o
gigante. Essa notícia espalhou-se rapidamente e foi parar no palácio
do rei. Davi tapou os ouvidos à voz dos embaixadores do caos; ele
não ouviu os profetas do pessimismo, nem inclinou os seus ouvidos
ao vozerio desolador da multidão apavorada.

Vencedores de gigantes não ouvem a voz dos pessimistas

Um vencedor de gigantes não segue o caminho dos medrosos. Os prognósticos agourentos dos covardes só infundem desespero. Eles olham para a vida com lentes cinzentas e só enxergam dificuldades. A voz da multidão quase sempre infunde medo e conduz ao fracasso. Aqueles que nunca mataram um gigante vão lhe dizer que os gigantes são invencíveis. Aqueles que correm até da sombra dirão que a estrada da vida está cheia de fantasmas. Aqueles que fracassam justificarão suas retiradas do campo de batalha.

Precisamos olhar não para a altura dos gigantes, mas por sobre seus ombros e agarrar a crise pelo pescoço. Precisamos olhar não para a carranca da crise, mas para as possibilidades que se ocultam por trás delas. A crise é um tempo de oportunidade, um terreno fértil de milagres. A crise é o cemitério dos covardes e a porta de entrada da terra prometida para os vencedores. Enquanto todos se assentam para chorar, os vencedores empunham as armas e triunfam. Enquanto os covardes só se encolhem de medo diante das dificuldades, os vencedores permanecem com os olhos fixos na recompensa e obcecados pela vitória. Por isso, não dê atenção aos medrosos. Você pode derrubar os gigantes e ser um vencedor.

O Salmo 137 apresenta-nos um quadro que elucida esse ponto. O povo de Israel havia sido banido de Jerusalém e levado para o exílio na Babilônia. Todos enfrentavam um passado de dor e perdas. Agora, eles estavam às margens dos rios da Babilônia, onde não desejavam estar, e fazendo o que não queriam. Como eles reagiram a essa situação?

Eles se entregaram à um estado de apatia coletiva. Eles se assentaram e dependuraram suas harpas. Em vez de cantarem suas canções no exílio, entregaram-se à nostalgia. Eles se assentaram para chorar e não para sonhar com novos tempos ou traçar estratégias para retornar à terra da promessa. Além disso, entregaram-se às memórias amargas e às lembranças de Sião. Eles estavam sem ânimo para viver o presente porque ainda moravam no passado. Deixaram de ser uma referência no presente porque estavam presos ao passado. Em vez de cantar em terra estranha, eles choraram. Em vez de abençoar o povo da Babilônia, sendo luz nas trevas, encheram o peito de ódio e clamaram por vingança. A única coisa que eles fizeram foi chorar pelo passado e desejar um futuro marcado de tragédia para os seus adversários. Disseram: *"Filha de Babilônia, que hás de ser destruída; feliz aquele que te der o pago do mal que nos fizeste! Feliz aquele que pegar teus filhos e esmagá-los contra a pedra"* (Sl 137.8,9).

Daniel e seus amigos estavam no mesmo cativeiro, mas escolheram outro caminho, outra atitude. Eles não esqueceram os dias passados, mas não deixaram de viver o presente e sonhar com o futuro. Eles decidiram influenciar o meio em que estavam vivendo. Superaram a crise e fizeram dela um instrumento de bênção para outras pessoas. Daniel tornou-se conhecido em toda a Babilônia, e mais do que isso, tornou seu Deus também conhecido. Ele se tornou maior que o rei da Babilônia. Caiu a Babilônia, mas Daniel permaneceu de pé.

As impossibilidades dos homens são possíveis para Deus. Um vencedor de gigantes mantém seus olhos em Deus e nunca nas circunstâncias. Pedro afundou no mar quando tirou os olhos de Jesus para colocá-los no fragor das ondas. Enquanto ele manteve os olhos em Jesus foi capaz de caminhar em meio à tormenta, mas no instante em que reparou na força do vento, teve medo e começou a ser engolido pelos vagalhões. Andar sobre as águas é algo impossível para o homem, mas quando nossos olhos estão fitos no Criador do céu e da terra, podemos experimentar o sobrenatural e pisar no solo firme dos milagres.

Um vencedor de gigantes não vê o passado como um obstáculo, mas enxerga o futuro como um campo fértil de gloriosas realizações. Juscelino Kubitscheck foi presidente do Brasil no período de 1956 a 1961. Seu pai era caixeiro-viajante e sua mãe professora. Ficou órfão de pai aos três anos. Teve uma infância muito pobre. Com o coração engravidado de sonhos, ainda muito jovem deixou a cidade de Diamantina e foi estudar Medicina em Belo Horizonte.

Lutava com tantas dificuldades para custear seus estudos que não lhe sobrava dinheiro nem para comprar uma cadeira, tendo que improvisar com um caixote de tomate. Mas esse homem venceu os seus gigantes. Concluiu o curso de Medicina e tornou-se um médico cirurgião, indo logo depois aperfeiçoar seus estudos na França e na Alemanha. Quando retornou ao Brasil, ingressou na política, elegendo-se deputado federal por duas legislaturas. Foi prefeito de Belo Horizonte, governador do Estado de Minas Gerais e por fim presidente da República, governando o Brasil de 1956 a 1961. Em 21 de abril de 1960 inaugurou a nova capital da República, Brasília, situada no coração do serrado brasileiro. Juscelino venceu a crise e tornou-se um ícone da nação.

História semelhante pode ser vista na vida do torneiro mecânico Luiz Inácio Lula da Silva, que saiu do sertão de Pernambuco para São Paulo, viajando em um pau-de-arara, em busca de novos horizontes. Timbrado pela pobreza, sem as portas abertas das oportunidades da educação, Lula foi um guerreiro destemido. Sua pobreza não sepultou os seus sonhos. Desde a juventude mergulhou seu coração nas tensões dos trabalhadores. Tornou-se líder de classes. Posicionou-se ao lado dos trabalhadores. Tornou-se a voz dos homens sujos de graxa. A cada passo que dava, os horizontes se alargavam. Até que despontou como líder, ganhou espaço, notoriedade, vez e voz. Seu sonho elasteceu. De líder sindical, passou a político de projeção. Galgou os degraus da popularidade, tornando-se deputado federal. Seus sonhos ainda foram mais ousados. Aspirou chegar à presidência da República, o posto mais alto da nação. Isto representava algo inédito: um plebeu, sem cruzar os umbrais de uma universidade, sem ostentar diplomas, pleiteava o posto de maior honra da nação. Em sua jornada rumo ao palácio do Planalto, sofreu três derrotas consecutivas, mas nunca se 3X abateu nem se acovardou. Até mesmo alguns de seus aliados políticos tentaram demovê-lo de concorrer ao último pleito e prosseguir em busca do seu sonho, mas ele bravamente resistiu, sem recuar diante das dificuldades.

Vencendo Gigantes

Muitos acharam que ele jamais conseguiria vencer os gigantes da política nacional. Mas sua perseverança triunfou sobre a voz dos pessimistas e Lula alcançou consagradora vitória, pondo os pés na rampa do Palácio do Planalto no dia 1 de janeiro de 2003 como presidente eleito da República Federativa do Brasil. Sua história pode ser controvertida, mas jamais apagada. Talvez ele tenha granjeado muitos inimigos nesse processo, mas ninguém pode negar que ele possui a marca dos vencedores de gigantes. Hoje Lula é considerado um ícone nacional, símbolo de um homem que, com otimismo inveterado, tapou os ouvidos à voz dos pessimistas para alcançar esplêndidas vitórias.

Vencedores de gigantes superam suas próprias limitações. A obsessão por vencer é maior do que suas próprias fraquezas. Eles alimentam-se de sonhos e não se abatem por causa dos pesadelos. Franklin Delano Roosevelt foi o único presidente dos Estados Unidos eleito para quatro mandatos consecutivos. Nascido no Estado de Nova York, em 30 de janeiro de 1882, Roosevelt estudou na Universidade de Harvard de 1900 a 1904, e formou-se em Direito pela Universidade de Colúmbia. Deixou a carreira de advogado para entrar na política, elegendo-se senador em 1910, pelo Estado de Nova York. Em 1920, foi derrotado como candidato à vice-presidência da República. No ano seguinte, quando estava com trinta e nove anos de idade, foi acometido de poliomielite e ficou paralisado da cintura para baixo. Em 1928 foi eleito governador do Estado de Nova York, cargo para o qual foi reeleito em 1930.

A despeito do seu problema físico, Roosevelt nutria no coração um sonho audacioso: ser presidente dos Estados Unidos. Os arautos do caos tentaram demovê-lo, mas ele, com determinação insuperável, lutou bravamente contra a enfermidade e não permitiu que ela se tornasse um gigante e o impedisse de realizar seus sonhos. Roosevelt não apenas superou seus próprios limites, mas alcançou alturas jamais conquistadas na saga da política americana. Foi eleito presidente dos Estados Unidos em 1932, realizando um mandato

Vencedores de gigantes não ouvem a voz dos pessimistas

esplêndido, tirando o país do atoleiro da crise financeira através de um programa econômico que ficou conhecido como *New Deal*. Foi reeleito em 1936, e novamente em 1940. Foi finalmente reeleito em novembro de 1944 mas, gravemente enfermo, não chegou a assistir à vitória aliada na Segunda Guerra Mundial, morrendo em 12 de abril de 1945, na cidade de Warm Springs, no Estado da Geórgia. Roosevelt conseguiu tirar seu país do mais profundo atoleiro financeiro e liderou a nação durante o dramático período da Segunda Guerra Mundial. Ele foi um vencedor de gigantes!

4

Vencedores de gigantes triunfam sobre as críticas

Ouvindo-o Eliabe, seu irmão mais velho, falar àqueles homens, acendeu-se-lhe a ira contra Davi, e disse: Por que desceste aqui? E a quem deixaste aquelas poucas ovelhas no deserto? Bem conheço a tua presunção e a tua maldade; desceste apenas para ver a peleja. Respondeu Davi: Que fiz eu agora? Fiz somente uma pergunta. Desviou-se dele [...] Porém Saul disse a Davi: contra o filisteu não poderás ir para pelejar com ele; pois tu és ainda moço, e ele, guerreiro desde a sua mocidade [....] Olhando o filisteu, e vendo a Davi, o desprezou (1 Sm 17.28-30, 33, 42).

Antes de vencer o gigante Golias, Davi precisou vencer aqueles que o criticavam.

Os críticos são como erva daninha, florescem em qualquer lugar. São os inimigos de plantão, que sempre estão à nossa espreita. Estão espalhados por toda parte, esperando para nos morder sem piedade. Eles estão dentro de casa, nas ruas, no trabalho, na escola e até na igreja.

Davi foi duramente criticado pelo seu próprio irmão Eliabe. A raiz da crítica era a inveja. Davi foi criticado também por Saul, que o julgou inepto e incapaz de enfrentar o gigante Golias. Saul subestimou Davi. Finalmente, Davi foi criticado pelo próprio duelista, o gigante Golias, que o desprezou, escarnecendo dele e do seu Deus. A motivação foi o sentimento de superioridade.

Você não pode ser um vencedor de gigantes sem antes vacinar-se contra o veneno dos críticos. O objetivo deles é sempre querer nos nivelar à sua mediocridade. Eles são medrosos e covardes e não toleram ver em nós uma atitude de confiança diante dos desafios da vida. Os críticos questionam nossas motivações, julgam nossos propósitos e assacam contra nós acusações pesadas e levianas, tentando macular a nossa honra e desbaratar nossos objetivos.

Os críticos são movidos pelo combustível da inveja. O invejoso é aquele que se sente infeliz por não ser como você e por não ter o que você tem. Caim matou Abel por inveja, em vez de seguir seu exemplo. Os irmãos de José tentaram se livrar dele por inveja, em vez de seguir os seus passos. Os fariseus, por inveja de Jesus, preferiram persegui-lo a seguir os seus ensinos. O invejoso é um egoísta inveterado, dono de uma auto-imagem doentia. Ele sofre de um avassalador complexo de inferioridade. É capaz de chorar com os que choram, mas jamais se alegrar com os que se alegram.

Algumas situações específicas nos tornam mais vulneráveis às críticas. Veremos a seguir quando as críticas podem nos machucar.

1. As críticas podem nos machucar quando vêm de alguém que deveria estar do nosso lado.

Eliabe era irmão de Davi. Ele era sangue do seu sangue. Como irmão mais velho, viu Davi crescer e despontar como músico de dotes excelentes. Ele conhecia o caráter piedoso de Davi e sabia que seu irmão era alguém escolhido por Deus. Mas Eliabe não se alegrou com o sucesso de Davi. A felicidade do seu irmão foi a sua tristeza. O sucesso de Davi foi o seu fracasso. A vitória de Davi foi a sua derrota. Há pessoas que não conseguem celebrar a vitória dos outros. Ficam tristes sempre que alguém é promovido. Sentem-se inseguras, ameaçadas e lesadas sempre que alguém é colocado num posto de preeminência. Eliabe irritou-se por causa da coragem de Davi. A coragem de Davi era uma denúncia e uma ameaça à sua covardia. Eliabe tinha pose, mas não tinha fibra; tinha tamanho, mas não caráter. Tinha cacoete de soldado, mas não a têmpera de um vencedor de gigantes.

2. As críticas podem nos machucar quando questionam nossas motivações.

Eliabe chamou Davi de presunçoso. Nada mais longe da verdade. A verdadeira motivação que fez Davi enfrentar o gigante foi o seu zelo pela glória de Deus. Os exércitos do Deus vivo estavam sendo afrontados e o nome de Deus escarnecido. As tropas de Israel estavam envergonhadas e humilhadas. Davi não podia aceitar que um filisteu incircunciso continuasse humilhando o povo de Deus. Por isso, se dispôs a enfrentar o temido duelista.

Os críticos são especialistas em torcer a verdade. Eles se alimentam da mentira. Eles nos julgam como se fôssemos iguais a eles. Por isso, questionam e torcem as nossas motivações. Eles não podem crer que haja pessoas honestas, com propósitos santos e puros. São pérfidos e destilam veneno em suas palavras. Agem como Satanás quando acusou Jó. Satanás questionou a motivação da fidelidade de Jó, argumentando que Jó só era fiel a Deus porque recebia muitas bênçãos de Deus. Satanás acusou Jó de ser uma pessoa hipócrita e interesseira. Para Satanás, Jó amava mais o dinheiro, a família e a saúde do que a Deus. Mas Jó não era semelhante a Satanás. Ele era filho de Deus. Possuía outro caráter, outros valores, outra motivação. Seu amor por Deus era superior a todas as outras devoções.

Eliabe não podia aceitar que Davi recebesse os louros daquela vitória. Afinal, Davi não passava de um simples pastor de ovelhas, enquanto que ele, Eliabe, era soldado e guerreiro. Os holofotes deveriam estar sobre ele, e não sobre Davi. Como Eliabe não tinha coragem suficiente para enfrentar Golias, ele colocou em dúvida a motivação de Davi.

3. As críticas podem nos machucar quando são contínuas.

Eliabe era um crítico inveterado, crônico e intermitente. Ele não dava pausa. Estava sempre buscando uma ocasião para atingir Davi. Em vez de lutar pelos seus próprios sonhos, buscava destruir os sonhos do irmão. Eliabe tinha o hábito de atacar Davi. As virtudes deste o incomodavam. Em vez de imitar Davi e seguir suas pegadas, ele tentava jogar lama em Davi. A resposta de Davi à sua insolente crítica mostra a personalidade doentia do irmão mais velho: "Que fiz eu agora?" Davi já estava acostumado com as palavras cheias de veneno de Eliabe.

Assim são aqueles que se alimentam do absinto da inveja, preferem destruir o outro a imitar suas virtudes. Davi era um jovem forte, corajoso, habilidoso e tinha uma profunda comunhão com Deus. Era um músico consagrado, um compositor inspirado. Mesmo trabalhando como pastor nas rudes e toscas montanhas de Belém, tinha grande sensibilidade ao dedilhar a harpa. Embora sendo o caçula da família, as virtudes de Davi lançavam sombras sobre a mediocridade de Eliabe, que tinha uma performance invejável, mas não um caráter firme. Só tinha aparência, mas sem vida. Em vez de aproveitar as oportunidades para amadurecer, resolveu atacar o irmão para impedir que ele se projetasse.

Mas as críticas não podem derrubar os vencedores de gigantes, ainda que sejam contínuas e cheias de veneno.

4. As críticas podem nos machucar quando vêm daqueles que nos conhecem.

Eliabe conhecia a Davi desde o seu nascimento. Sabia de sua devoção a Deus, de sua coragem e força, mas sobretudo, sabia que Deus já o havia escolhido para ser o futuro rei de Israel. Eliabe criticou Davi porque queria estar no lugar dele, apesar de não ter o caráter de Davi. A crítica que machuca não é a que procede de pessoas estranhas, mas daquelas mais chegadas. Quando somos atacados por pessoas próximas a nós, que convivem conosco e nos conhecem, isso fere a nossa alma.

Davi não discutiu com Eliabe, ele simplesmente saiu dali. A atitude de Davi revela domínio próprio. A primeira vitória de Davi foi sobre si mesmo diante de seu irmão. Não precisamos fazer o jogo daqueles que querem paralisar os nossos passos. Vencedores de gigantes não têm que se defender das críticas ruidosas dos covardes nem perder tempo com os invejosos, mas caminhar resolutamente em direção à vitória. Fuja daqueles que gratuitamente buscam humilhá-lo ou transferir para você a sua covardia, projetando em você a sua mediocridade. Não desperdice seu tempo nem suas emoções com aqueles que se intrometem na sua vida e tentam afastá-lo do caminho da vitória.

5. As críticas podem nos machucar quando demonstram ingratidão.

Davi não foi até o campo de batalha por iniciativa própria. Ele estava ali porque Jessé, seu pai, assim lhe ordenara. Seu propósito não era diminuir seus irmãos, Eliabe, Abinadabe e Samá, que estavam na guerra, nem tirar-lhes o brilho da honrosa posição, mas apenas alimentá-los. Davi foi até lá levar comida para os irmãos (1 Sm 17.17). A motivação de Davi não era presunção nem maldade, como Eliabe o acusou (1 Sm 17.18), mas humildade e obediência, abnegação e prontidão para servir. Davi foi até o campo para abençoá-los e não para afligi-los. A acusação feita a Davi era um desatino, uma inversão da verdade, uma clamorosa injustiça, uma consumada ingratidão.

Davi não estava no campo de batalha motivado por presunção ou maldade. Ele foi até lá para saber se os seus irmãos estavam bem (1 Sm 17.22). Era o bem, e não o mal de seus irmãos que Davi buscava. Davi estava preocupado com o bem-estar físico e emocional dos seus irmãos, mas Eliabe atirou contra ele uma seta envenenada de ingratidão.

As críticas injustas e recheadas de ingratidão nos machucam. As pessoas são capazes de ferir as mãos que as abençoam. Elas são capazes de questionar nossas motivações mais puras e nossos gestos mais sublimes. Vencedores de gigantes não se deixam abater com as críticas, não dependem de elogios nem desanimam com as críticas.

6. As críticas podem nos machucar quando vêm cheias de descontrole emocional.

Quando Eliabe ouviu Davi falando com os homens de Israel acerca da sua disposição de enfrentar o gigante, acendeu-se-lhe a ira contra Davi (1 Sm 17.28). A ira descontrolada provoca grandes tragédias. Um homem iracundo, sem domínio próprio, é um barril de pólvora, uma mina prestes a explodir, um perigo àqueles que estão à sua volta. A ira descontrolada é uma porta aberta para a ação devastadora do diabo (Ef 4.26,27). Aqueles que não controlam seus sentimentos, não controlam a língua. E a língua movida por um coração irado é um fogo incontrolável, um veneno mortífero, um mundo de iniqüidade.

Vencedores de gigantes são alvos das críticas mais duras. Eles não apenas derrubam os gigantes, eles incomodam os covardes. Suas vitórias representam a derrota dos invejosos. Seu triunfo tem um gosto amargo para aqueles que não toleram ver o seu sucesso.

Uma antiga lenda ilustra bem esse fato. Certa vez, uma cobra estava perseguindo um vagalume. Depois de três dias de implacável perseguição, o vagalume parou e perguntou à asquerosa víbora: "Posso lhe fazer três perguntas?" A cobra respondeu: "Não costumo fazer concessão, mas vamos lá. O que deseja saber?" O vagalume então perguntou: "Eu lhe fiz algum mal?" "Não", respondeu a víbora. "Faço parte da sua cadeia alimentar?" "Também não", respondeu a serpente. "Por que, então, você quer acabar comigo?" A cobra respondeu: "Porque não tolero ver você brilhando". Assim são os críticos invejosos, movidos pelo combústivel da ira.

7. As críticas podem nos machucar quando visam nos humilhar.

Eliabe acrescentou um detalhe à sua crítica a Davi que revelou sua sórdida intenção de humilhá-lo: *"Por que desceste aqui? E a quem deixaste aquelas poucas ovelhas no deserto?"* (1 Sm 17.28). Sua intenção evidente era afirmar que Davi não tinha cancha para ser um soldado, que não passava de um insignificante e pobre pastor, com poucas ovelhas. As palavras de Eliabe eram carregadas de veneno. Ele deixava vazar de seu coração invejoso sua fraqueza, insegurança, e complexo de inferioridade. Pretendia se sair bem diminuindo Davi; aparentar ser maior, humilhando seu irmão.

Vencedores de gigantes não precisam de aplausos nem perdem o entusiasmo quando são humilhados. A grandeza de Davi não estava na riqueza material, mas na dignidade do seu caráter. Era um simples pastor, mas conhecia o grande Deus. Era pobre, mas possuía tudo. Vivia na solidão do deserto e não nas frentes de batalha, mas tinha coragem e disposição para vencer os gigantes!

8. As críticas podem nos machucar quando vêm de pessoas que não acreditam em nosso potencial.

Saul estava com muito medo do gigante (1Sm 17.11) e subestimou a capacidade de Davi para enfrentá-lo. O crítico sempre está pronto a achar defeito em tudo que os outros fazem, mas ele mesmo não se dispõe a fazer alguma coisa. O crítico é aquela pessoa que polui todas as soluções que você encontra. Aquele que nunca matou um gigante irá lhe dizer que isso é impossível. Saul disse que Davi era muito jovem e também inexperiente (1 Sm 17.31-33) para enfrentar um gigante adestrado nas batalhas como Golias. A mensagem que Saul estava transmitindo a Davi era: "Veja bem, ele é forte, e você é fraco. Ele é um guerreiro, e você um simples pastor de ovelhas. A luta é desigual e você não tem nenhuma chance de prevalecer".

Os medrosos sempre tentam transferir aos outros o seu pessimismo. Eles tentam nivelar as pessoas à sua própria mediocridade. O que eles sabem fazer é contar os inimigos e fugir na hora da peleja. Davi viu primeiro o potencial, depois o problema. O exército de Saul viu primeiro o problema, depois o potencial. O exército viu Golias, Davi viu a Deus. Oswald Sanders disse que os olhos que olham são comuns, mas os olhos que vêem são raros. Foi por isso que Eliseu orou para que Deus abrisse os olhos do seu moço, a fim de que ele visse a majestade dos exércitos de Deus e não a opulência dos adversários.

Vencedores de gigantes não dão ouvidos aos arautos do pessimismo. Eles não olham para as dificuldades, mas para as possibilidades. Eles não gastam tempo medindo a altura do inimigo, mas fortalecem os braços para vencê-lo. Eles não fogem assustados com o rugido do inimigo, mas correm para a linha de frente, para a zona de combate, porque estão obcecados pela vitória.

Saul enxergou três empecilhos em Davi:

Primeiro, ele considerou Davi incapaz – Saul disse: "Você não pode...". Nunca diga para uma pessoa que ela não pode. Aquele que confia em Deus pode o impossível. O apóstolo Paulo, mesmo preso, na ante-sala do martírio, disse: *"Tudo posso naquele que me fortalece"* (Fp 4.13).

Segundo, ele considerou-o inexperiente – Saul disse: "Você não tem experiência". O sucesso de uma pessoa não está na sua idade cronológica, mas na sua confiança em Deus. Há jovens sábios e velhos tolos. Há jovens cansados e velhos cheios de sonhos e projetos. Davi nunca tinha estado numa guerra, mas tinha intimidade com o Senhor dos Exércitos. Ele nunca tinha empunhado uma espada, mas tinha matado um urso e agarrado um leão pela barba e vencido. Davi nunca tinha usado uma armadura nem empunhado um escudo, mas manejava com perícia sua funda. Ele não tinha performance para fazer parte do exército de Saul, mas tinha determinação para enfrentar e vencer o gigante.

Terceiro, ele considerou-o muito jovem – Saul disse: "Você é ainda muito jovem". Davi era jovem, mas corajoso. Era jovem, mas confiava em Deus e possuía a fibra de um vencedor de gigantes. Os jovens também podem ser heróis e vencedores. Eles podem triunfar sobre os gigantes. Se você é jovem, não permita que os críticos contaminem seu coração e deixem você desanimado por causa da sua pouca idade. Eu saí da casa dos meus pais com doze anos de idade para estudar. Ainda criança deixei o campo, a lavoura, a família para enfrentar os desafios da cidade grande e descortinar o futuro. As lutas foram muitas, mas um a um os gigantes foram sendo derrotados. Hoje sei que Deus dá vitória àqueles que esperam nele.

9. As críticas podem nos machucar quando vêm daqueles que querem nos destruir.

Golias não só desafiou os exércitos de Israel, mas também insultou Davi (1 Sm 17.34-36). Ele era arrogante, insolente e blasfemo. Os gigantes têm a capacidade de nos intimidar quando insistem em nos desafiar. Eles se tornam ameaçadores quando têm uma certa reputação e insistem em aparecer. Eles se tornam perigosos quando nos derrotam psicologicamente e apavoram os que estão do nosso

Vencendo Gigantes

lado. Ficamos fragilizados quando tudo o que conseguimos fazer é nos reunir, mas não enfrentamos o gigante, ou quando nossos líderes encharcam-se de medo e fogem da peleja.

Golias usou três métodos diferentes para humilhar os seus opositores:

Primeiro, ele afrontou as tropas de Israel (1 Sm 17.10,36) – Os soldados de Saul fugiram porque olharam para a bravura, para a altivez e insolência do gigante. Os soldados olharam para si mesmos e viam-se como um bando de homens medrosos. Mas Davi viu a cena com outros olhos. Para ele, Golias não estava apenas afrontando os soldados de Saul, mas afrontando os exércitos do Deus vivo (1 Sm 17.36). A batalha não era apenas no plano físico, mas sobretudo, na dimensão espiritual. Golias não estava se insurgindo apenas contra Israel, mas contra o Deus de Israel. Assim, a diferença entre os vencedores de gigantes e aqueles que fogem dos gigantes é simplesmente uma questão de visão e perspectiva. Precisamos saber que mais são aqueles que estão conosco do que aqueles que estão contra nós. Precisamos saber que se Deus é por nós, ninguém poderá ser contra nós. Se Deus está conosco, somos maioria absoluta. Somos invencíveis. Podemos derrubar gigantes.

Segundo, ele amaldiçoou Davi (1 Sm 17.43) – Os gigantes têm uma boca insolente e cheia de maldição. Eles nos desprezam e invocam sobre nós a maldição de seus deuses. Eles tentam nos assustar e nos vencer pelo seu aparato bélico e pelas suas divindades ameaçadoras. Mas os vencedores de gigantes não entram nessa peleja confiados no braço da carne. A vitória não é dos fortes nem dos que correm melhor, mas daqueles que esperam no Senhor. Davi disse: *"Eu vou contra ti em nome do Senhor dos Exércitos, o Deus dos exércitos de Israel, a quem tens afrontado"* (1 Sm 17.45). A afronta de Golias não era contra os homens, mas contra Deus. Ninguém pode lutar contra Deus e prevalecer. Ninguém pode desafiar o Deus Todo-poderoso e escapar. Davi enfrentou o gigante não com a possibilidade da vitória, mas com a certeza da vitória, porque Deus jamais saiu vencido de uma batalha.

Terceiro, ele ameaçou acabar com Davi (1 Sm 17.44) – Os gigantes pensam que são invencíveis. Eles jamais contam com a derrota. Eles são convencidos. Eles sofrem de um grande complexo de inexpugnabilidade. São megalomaníacos. Mas quem se exalta será humilhado. Quem tenta ficar de pé estribado no bordão da autoconfiança precisa beijar o chão e se alimentar com o pó da derrota. A maldição imprecada a Davi voltou-se contra Golias. Ele mesmo sofreu o golpe fatal de suas medonhas previsões. Ao afrontar Davi, ele estava afrontando o Deus de Davi. Quem toca nos filhos de Deus toca na menina dos olhos de Deus. Quem declara guerra contra os filhos de Deus chama o próprio Deus Todo-poderoso para a briga. Naquele combate, o gigante caiu. Davi, com uma só pedra, jogou o gigante ao chão. A cabeça de Golias foi cortada pela sua própria espada. Ele pereceu através de suas próprias armas. Golias morreu. Davi prevaleceu. Israel ficou livre e o nome de Deus foi glorificado. Davi não se intimidou porque tinha paixão em promover a glória de Deus; tinha seus olhos na recompensa e sabia que o próprio Deus é quem estava pelejando por ele.

Os tempos mudaram. Você mudou. Os gigantes se travestiram e usam novas roupagens, mas eles ainda rondam a nossa vida e nos espreitam com insolentes ameaças. Muitos ainda estão com medo, dando as costas ao inimigo e fugindo acovardados. Muitos estão desacreditados aos seus próprios olhos, enfileirados apenas para fugir ao sonido do próximo estardalhaço do gigante. Volte seus olhos para Deus em vez de inclinar os seus ouvidos às bravatas dos gigantes. Você não foi criado para viver debaixo das botas dos gigantes. Ponha sua fé naquele que está assentado no alto e sublime trono. Enfrente o seu gigante em nome do Senhor dos Exércitos. Não deixe que a insolência dos gigantes tire de você a certeza da vitória e a pressa para vencer.

A revista *Time* de abril de 1986 publicou um artigo sobre algumas pessoas que foram consideradas sem perspectiva de futuro. Todas foram vistas pelos críticos como condenadas ao fracasso. Quer saber os nomes de algumas dessas pessoas?

Beethoven – Seu professor certa feita fez o seguinte comentário sobre ele: "Esse jovem tem uma maneira estranha de manusear o violino. Prefere tocar suas próprias composições ao invés de aprimorar sua técnica". Esse mesmo professor avaliou que não havia esperança para ele como compositor. Bethoven não aceitou esse prognóstico pessimista a seu respeito. Ele superou todas as dificuldades e triunfou. Mesmo tendo ficado completamente surdo, ele se tornou um dos músicos mais excelentes e consagrados de toda a história. Seus críticos tiveram que engolir suas previsões negativas e ver o sucesso desse gigante da música clássica. Mesmo carregando o fardo da surdez, compôs músicas timbradas pelo mais alto grau de excelência.

Walt Disney – Foi despedido pelo editor de um jornal sob alegação de "falta de idéias e de criatividade". Walt Disney foi à falência várias vezes, mas em vez de naufragar na tempestade das circunstâncias adversas e sucumbir pela impiedade dos críticos, triunfou e hoje seu nome é conhecido no mundo inteiro como o criador do maior parque de diversões do planeta. Amealhou riqueza, fama e prestígio, deixando seus críticos com o sabor amargo da derrota.

Thomas Alva Edison – Sua professora o devolveu à mãe por considerá-lo inapto para o aprendizado. Sua mãe, sem perder o ânimo, resolveu investir em sua educação. Certa feita, ele perguntou à sua mãe por que a escola o recusara, e ela lhe respondeu: "Porque você é muito inteligente e a escola não consegue acompanhar seu ritmo de aprendizado". Thomas Alva Edison superou as dificuldades e galgou o honroso posto de maior cientista do mundo, tendo registrado mais de mil invenções, entre elas a lâmpada elétrica. No dia de sua morte, como forma de homenageá-lo, desligaram todas as luzes por alguns minutos, mostrando como seria o mundo sem ele.

Albert Einstein – Sua tese de doutorado em Bonn foi considerada irrelevante e sofisticada. Alguns anos depois, foi expulso da Escola Politécnica de Zurich. Alguns de seus críticos acharam que não havia futuro para ele, mas Einstein galgou as alturas excelsas, fazendo vôos altaneiros em suas conquistas científicas. Tornou-se mundialmente conhecido por ter descoberto a teoria da relatividade, que lhe valeu o Prêmio Nobel de Física.

Luiz Pasteur – Foi um estudante medíocre. Em uma turma de vinte e dois alunos de química, obteve apenas a décima quinta colocação. Mas devemos a ele a descoberta da penicilina, que tanto tem contribuído para o tratamento dos males que afligem a humanidade.

Henry Ford - Foi o pioneiro na fabricação de carros em série. Foi à falência cinco vezes antes de ser bem-sucedido nos negócios.

Não permita que os críticos roubem as suas forças ou tirem de você a sede da vitória. Enfrente os gigantes e vença-os!

5

Vencedores de gigantes não usam armas alheias

Saul vestiu a Davi da sua armadura, e lhe pôs sobre a cabeça um capacete de bronze, e o vestiu de uma couraça. Davi cingiu a espada sobre a armadura e experimentou andar, pois jamais a havia usado; então, disse Davi a Saul: Não posso andar com isto, pois nunca o usei. E Davi tirou aquilo de sobre si. Tomou o seu cajado na mão, escolheu para si cinco pedras lisas do ribeiro, e as pôs no alforje de pastor, que trazia, a saber, no surrão, e, lançando mão da sua funda, foi-se chegando ao filisteu (1 Sm 17. 38-40).

Antes de enfrentar o gigante Golias, Davi enfrentou o gigante institucional. Havia falta de liderança em Israel. Saul estava com medo. Abner, o comandante das tropas, não dava notícia. Os soldados ainda estavam tremendo depois da última aparição de Golias. Faltavam homens de coragem para enfrentar a crise.

Saul achava que já que Davi estava disposto a enfrentar o gigante, tinha que fazê-lo do seu jeito, por isso, colocou a sua armadura em Davi, mas em vez de ajudá-lo, isto se tornou um estorvo. Então, Davi disse: *"Não posso andar com isto, pois nunca o usei. E Davi tirou aquilo de sobre si"* (1 Sm 17.39).

Veremos a seguir algumas importantes lições que esse episódio nos ensina.

1. Seja autêntico, você é uma pessoa singular.

Não podemos enfrentar os gigantes usando armas alheias. Os gigantes não se assustam com o nosso aparato nem cedem pelo fato de estarmos bem equipados. A armadura de Saul não serviu para Davi. Eles eram pessoas diferentes. Saul era um homem de grande estatura, o mais alto de todo o reino. Sua armadura não servia para outras pessoas, muito menos para Davi. Vestir roupa de rei sem ser rei não ajuda a enfrentar os gigantes. Certamente, a armadura de Saul não podia ficar bem ajustada em Davi. Era uma bagagem extra, um peso inútil, um verdadeiro estorvo. Por isso, Davi disse: "Não posso andar com isto", e tirou a armadura de sobre si.

Davi não podia nem andar, muito menos lutar com a armadura de Saul, que era absolutamente inadequada para ele. Por isso, logo se desvencilhou dela. Precisamos ter coragem de remover da nossa vida tudo aquilo que é peso morto e bagagem inútil. Se queremos vencer os gigantes não podemos usar uma armadura que não se ajusta a nós nem ostentar armas que não sabemos manusear com perícia.

A armadura de Saul seria apenas uma máscara para Davi. Saul queria que Davi se igualasse a ele, um homem medroso e covarde. Saul queria que Davi fosse mais um dos seus fracassados soldados. Não adianta tentar ser aquilo que você não é. As máscaras podem nos dar uma sensação momentânea de segurança, mas elas não nos abrem caminho para a vitória. Elas podem nos esconder apenas por um breve tempo, mas não são seguras, e geralmente caem nas horas mais impróprias.

Davi não se preocupou em agradar ao rei usando uma armadura que sabia ser inútil. Vencedores de gigantes não "fazem média", não fingem, não são hipócritas nem procuram agradar. Suas atitudes são coerentes. Integridade inegociável é a marca dos vencedores. Durante as crises, as pessoas esperam que você se torne igual a elas. Seja você mesmo. Seja autêntico. Você é uma pessoa única e singular. Você pode vencer os gigantes com sua própria habilidade e com as armas que Deus lhe confiou. Você pode vencer no mesmo campo juncado de perdedores. Tire a armadura de Saul e avance contra os gigantes com as armas que Deus colocou em suas mãos.

2 Tenha coragem de ser diferente, você não precisa ser igual aos outros.

Muitos líderes surgem no vácuo daqueles que fracassaram. Davi ousou ser diferente e fazer diferença. Ele trocou a armadura de Saul pela funda do pastor. Ele deixou o convencional para manejar uma arma inédita. Não use alguma coisa só porque todo o mundo está usando. Não se massifique nem se iguale à maioria medíocre. Você é uma pessoa especial, levantada por Deus para uma obra especial. Sua vocação é vencer. Os mesmos gigantes que apavoram os covardes e os fazem fugir de medo tombarão diante da sua coragem e beijarão o pó quando você empunhar as armas que Deus lhe confiou.

Os vencedores não são unanimidade. Eles são solitários, distinguem-se da maioria porque têm coragem de ser diferentes. Quando os soldados de Saul olharam para Golias eles fugiram, tomados de medo. Durante quarenta dias, de manhã e à tarde, Golias fez com que os soldados de Israel ensarilhassem as armas e fugissem. Davi viu o gigante pela primeira vez e dispôs-se a enfrentá-lo em nome do Senhor dos Exércitos. A coragem de Davi destacou-se em meio a um acampamento coberto de covardia. Os vencedores não olham para as circunstâncias com os olhos da maioria. Eles vêem as mesmas coisas, mas sob outra perspectiva. Aquilo que para os outros é ameaça, para os vencedores é oportunidade.

O uso de armadura era uma regra convencional nos duelos. Todos usavam armaduras nos duelos. Mas Davi tinha outros métodos e outras armas. Nem Saul, nem os soldados de Israel, nem mesmo o gigante Golias podiam avaliar a coragem e os recursos que Davi possuía. Ele avançou para a linha de frente da batalha sem a proteção de uma pesada armadura e sem empunhar uma espada afiada. A única coisa que ele levava nas mãos era uma funda. Dentro do seu alforje de pastor havia cinco pedrinhas, verdadeiros torpedos, mísseis rigorosamente endereçados para atingir a testa do gigante. Davi sabia manejar sua funda com perícia, por isso não a trocou pela espada. Ele teve ousadia suficiente para não seguir a orientação do rei nem se igualar aos seus tímidos soldados. Ele teve coragem de ser diferente.

Muitas pessoas fracassam porque vivem no palco da vida como atores, fingindo ser aquilo que não são. Desempenham um papel, fingindo ser diferentes do que são na vida real. Imitam os derrotados em vez de correr para a vitória. A Bíblia registra a história de Sansão. Sansão foi juiz em Israel. Ele era um jovem forte e um guerreiro imbatível. Mas a força de seus braços não era sustentada pela fortaleza do seu caráter. Seu nome significa "pequeno sol" e sua missão era brilhar num tempo de trevas. Mas Sansão fracassou porque não teve coragem de ser diferente. Ele foi capaz de vencer uma multidão, mas não se mostrou capaz de dominar suas próprias paixões. Com uma queixada de jumento foi capaz de matar mil filisteus, mas não foi capaz de livrar o seu coração dos pecados dos filisteus. Foi capaz de arrancar os portões de Gaza e levá-los sobre os ombros, mas não foi capaz de livrar-se dos pecados de Gaza. Por ser nazireu, ele não podia tocar em cadáver, mas insensatamente foi buscar na caveira de um leão morto um favo de mel. Foi buscar prazer e doçura na podridão. Como nazireu, não podia beber vinho, mas na festa do seu malfadado casamento ofereceu um banquete que durou uma semana, imitando os jovens da época. Ele não teve coragem de ser diferente. A maior força de uma pessoa é ter a ousadia de dizer "não" quando todos dizem "sim". A maior coragem de uma pessoa é não negociar os seus valores, convicções e princípios mesmo em meio a uma geração corrompida. Como nazireu, Sansão não podia cortar o cabelo, mas ao deitar no colo de uma víbora, não apenas seu cabelo foi cortado, mas também seus olhos foram vazados. De juiz a escravo. De pequeno sol à escuridão. De homem controlado por Deus a homem dominado pelo inimigo. Tudo porque ele não teve coragem de ser diferente, de ser autêntico. Estejamos alertas!

3. Especialize-se naquilo que você faz, destaque-se da maioria.

Toda pessoa que nunca matou um gigante vai lhe dizer que isso é impossível. Mas você não tem que ser igual a todo o mundo. Você é uma pessoa única, singular, especial. Você pode alcançar aquilo que ninguém ainda alcançou. Você pode conquistar vitórias inéditas. Há muitas descobertas a serem feitas, muitas vitórias ainda a serem alcançadas, muitos gigantes a serem derrotados.

Davi era um pastor de ovelhas e para defender o seu rebanho agarrou um leão pela barba e matou um urso. Ele ia às últimas conseqüências para fazer o seu trabalho com o selo da excelência. Destruidores de gigantes são construídos pelos sucessos do passado (1 Sm 17.34-37). John Maxwell afirmou que nós precisamos de vitórias não para celebrar, mas para nos elevar. Davi venceu um urso e um leão. Ele era um homem experimentado. Ele tinha experiência da proteção de Deus. Não vivia numa estufa, numa re-doma de vidro. Ele conhecia a fidelidade de Deus nas lutas renhidas do cotidiano, e as experiências do passado garantiam a vitória no futuro. O Deus que agiu no passado continua agindo no presente. A imutabilidade de Deus é o penhor da nossa vitória e a causa de não sermos destruídos.

Não importa o que você faz, o que importa é o quanto você se dedica àquilo que faz. Torne-se um especialista na sua área de atividade. Os medíocres não se esforçam. Eles não são criativos. Eles vivem fugindo, dando desculpas e justificando seus fracassos. Eles são preguiçosos, eles não se afatigam, não calejam as mãos nem investem na preparação, por isso não podem vencer os gigantes.

O mundo de hoje requer especialistas, pessoas peritas no que fazem. Quando alguém precisa de um médico, busca sempre o melhor. Quando precisa de um engenheiro, procura conhecer os projetos que já executou. Quando procura uma escola para os filhos, preocupa-se em conhecer sua filosofia de ensino e seu compromisso com a excelência. Quando busca uma igreja, procura saber se o pastor é um homem honrado, íntegro, que prega a Palavra com fidelidade.

O segredo do sucesso é não se conformar com a mediocridade. Davi buscava o padrão de excelência em tudo que realizava. Como músico, era um prodígio. Como compositor, um homem inspirado por Deus. Como pastor, um homem valente, disposto a dar a vida para salvar suas ovelhas. Como guerreiro, um soldado destemido capaz de enfrentar o insolente gigante.

Davi não era um aventureiro inconseqüente. Ele não tinha apenas coragem, mas também preparo. Com uma funda e cinco pedrinhas ele foi mais eficiente que todo o exército de Saul, aparelhado com armaduras e espadas. Ele correu para a linha de frente não porque tinha a síndrome de mártir, mas porque tinha a marca de um vencedor. Apanhou cinco pedrinhas e venceu com a primeira, porque estava determinado a vencer. Caso a primeira pedra falhasse, ele não pretendia recuar. A perseverança na luta é o distintivo dos campeões.

Precisamos nos especializar naquilo que fazemos. Henry Ford foi o pioneiro na fabricação de carros em série no mundo. Charles Steinmetz era anão e possuía sérias deficiências físicas, mas era um engenheiro brilhante, considerado uma das maiores inteligências do mundo na área da eletricidade. Foi ele quem fabricou os primeiros geradores para a fábrica da *Ford*, em Dearborn, Michigan. Um dia, os geradores queimaram, e toda a fábrica parou de funcionar. Mandaram chamar vários técnicos e eletricistas para consertá-los o mais rápido possível, pois a empresa estava perdendo muito dinheiro. Mas nenhum deles pôde consertá-los. Então, Henry Ford mandou chamar Charles Steinmetz. O gênio chegou à fábrica, passou algumas horas mexendo nos motores, depois ligou a chave geral, e a fábrica inteira voltou a funcionar. Alguns dias mais tarde, Henry Ford recebeu a conta pelos serviços prestados por Steinmetz. O valor? Dez mil dólares! Embora Ford fosse muito rico, devolveu a conta com um bilhete: "Charles, essa conta não está muito alta para um serviço de poucas horas, em que você apenas deu uma mexida nos motores?". Steinmetz devolveu a conta para Ford, desta vez discriminando os valores: "Quantia a ser paga por mexer nos motores: dez dólares. Quantia a ser paga por saber o lugar certo onde mexer: nove mil e novecentos e noventa dólares. Total pelos serviços prestados: dez mil dólares". Ford pagou a conta.

Você é um especialista naquilo que faz? Hoje você é melhor do que foi ontem? Você está florescendo onde foi plantado? Você é um vencedor de gigantes?

6

Vencedores de gigantes são determinados a vencer

Sucedeu que, dispondo-se o filisteu a encontrar-se com Davi, este se apressou e, deixando as suas fileiras, correu de encontro ao filisteu (1 Sm 17.48).

Antes de vencer o gigante à sua frente, Davi precisou triunfar sobre os opositores ao seu redor. Há assassinos de sonhos que nos espreitam e tentam nos afastar da vitória. Há aqueles que nos desprezam e também nos subestimam. Não permita que esses assassinos de sonhos afastem você do caminho do sucesso.

Antes de derrubar o gigante Golias, Davi precisou vencer outros obstáculos. Antes de você subir ao pódio, precisa conquistar batalhas menores. Davi não se sentiu humilhado por ter ficado no campo cuidando das ovelhas em vez de ser convocado para a guerra. Ele não se sentiu humilhado quando seu pai o enviou, como uma espécie de *office boy*, para levar comida a seus irmãos no campo de batalha. Ele não se entregou à vaidade, mesmo sabendo que o plano de Deus era colocá-lo no trono, em lugar de Saul. Os vencedores nunca começam como destruidores de gigantes. Por Davi ter cumprido com fidelidade as tarefas menores, ele agora estava no centro do palco, sob as luzes da ribalta, travando uma decisiva batalha, que iria consagrá-lo como o maior herói nacional.

O primeiro passo para resolver qualquer problema é começar. Davi não andou para a linha de batalha, ele correu. Estava ansioso para ganhar, para vencer. Ele era um homem determinado a vencer, e venceu. Esse episódio nos ensina algumas importantes lições:

1. Em vez de correr do inimigo, avance contra ele.

Davi não esperou que Golias o atacasse. Ele não era homem de retaguarda, mas de vanguarda. Ele não se intimidou com as insolências e bravatas do gigante. Não ficou na retranca, acuado. Ele tomou a iniciativa, correu, avançou e deu o primeiro passo para a vitória consagradora. O mesmo Davi que já havia vencido o desânimo da multidão, as críticas invejosas de Eliabe, o preconceito de Saul, agora vencia o escárnio de Golias.

A vida é um campo de batalha. A nossa luta não é contra carne e sangue. O diabo e suas hostes infernais são os nossos arquiinimigos. Não somos chamados para contar os inimigos nem para temê-los. Jesus disse que as portas do inferno não prevalecerão contra a igreja. Não é o inferno que avança contra a igreja para acuá-la, mas a igreja que avança contra o inferno para desbaratá-lo. Não corremos do inimigo, avançamos contra ele. O nosso inimigo já foi despojado na cruz. Jesus já lhe tirou a armadura. Suas insolentes maldições não podem mais nos intimidar. Maior é aquele que está em nós. Deus não nos ordena a temer o inimigo, mas a resistir e vencer. A igreja é um exército com bandeiras, um povo vencedor e vitorioso. Devemos seguir as pegadas daquele que saiu para vencer e venceu!

2. Em vez de se intimidar com as ameaças e bravatas do inimigo, vença-o.

Golias subestimou e desprezou Davi, fazendo pouco caso dele. O gigante o amaldiçoou, invocando contra ele a maldição dos seus deuses iracundos. Ameaçou-o, prometendo dar a sua carne às aves do céu (1 Sm 17. 41-44). Mas Davi não deu ouvidos à insolência do inimigo. Ele sabia quem era Golias e quem estava com ele. Toda a maldição que Golias lançou contra Davi caiu sobre a sua própria cabeça. Em vez dos deuses filisteus tripudiarem contra Davi, foi o Deus de Davi, o Senhor dos Exércitos, que se tornou conhecido em toda a terra (1 Sm 17.45,46) por causa da retumbante vitória do pastor de Belém. Davi sabia que sua vitória não era apenas uma questão de esforço humano, mas sobretudo, fruto da intervenção divina (1 Sm 17.47).

Há muitas pessoas que vivem assustadas com medo das ameaças dos espíritos malignos. Vivem no tronco do adversário, prisioneiras do medo. São chantageadas e não conseguem se libertar do regime de terror. Os espíritos das trevas ameaçam com vingança aqueles que tentam escapar de suas mãos asquerosas. Mas a fúria e a maldição do inimigo não podem alcançar aqueles que estão nas mãos de Deus, sob as asas do Onipotente. A esses, o maligno não toca. O diabo ruge como leão, mas ele não é leão. O verdadeiro leão, o Leão da tribo de Judá, o vencedor invicto de todas as pelejas, é o Senhor Jesus, aquele que esmagou a cabeça da serpente, triunfou sobre a morte e nos garantiu a vitória. Quem anda com Jesus é mais que vencedor.

3. Em vez de temer a derrota, não se contente com nada menos que a vitória.

Davi não fez nenhuma provisão ou previsão para a derrota. Seu lema era vencer ou vencer. Na sua agenda não havia espaço para o fracasso. Derrota era uma palavra que não constava de seu dicionário.

Por essa razão, ele colocou cinco pedras em seu alforje de pastor. Se a primeira pedra falhasse, ele ainda teria mais arsenal e mais munição. A luta poderia ser longa, mas a vitória era certa. Deus não nos promete ausência de lutas, mas nos garante a vitória; a caminhada não é fácil, mas a chegada é segura. Vencedores de gigantes não desistem diante da primeira dificuldade. Eles não se contentam com nada menos do que a vitória.

Vencedores de gigantes transformam dificuldades em pontes para vitórias mais consagradoras. O maior presidente americano de todos os tempos, Abraham Lincoln, nasceu no Estado de Kentucky, em 12 de fevereiro de 1809. Filho de lavradores, desde cedo teve de trabalhar arduamente na lavoura. Perdeu a mãe. muito cedo. A duras penas conseguiu estudar e obter seu diploma em Direito. Desejoso de progredir, o jovem Lincoln pedia livros emprestados a amigos e vizinhos para ler depois das suas atividades diárias. Estudava até ficar exausto, varando as noites debruçado sobre os livros. Entrou na política, tornando-se um dos homens de maior destaque da história dos Estados Unidos. Foi eleito o décimo sexto presidente da nação e seu nome tornou-se famoso no mundo inteiro, dando nome a ruas, avenidas, bancos, universidades e automóveis. Foi um dos grandes paladinos na luta pela abolição da escravatura nos Estados Unidos. Enfrentou com galhardia o dramático período da guerra civil americana. Lincoln foi um homem temente a Deus, talhado para grandes embates, obcecado pela vitória. Ele foi um vencedor de gigantes.

Suas vitórias não foram fáceis. Sua marca foi a perseverança. Ele transformou suas derrotas temporárias em degraus para alcançar vitórias mais consagradoras. Aos vinte e dois anos enfrentou um fracasso nos negócios. Aos vinte e três anos foi derrotado ao pleitear um cargo legislativo. Aos vinte e quatro anos, fracassou novamente nos negócios. Aos vinte e cinco anos conseguiu se eleger para um cargo legislativo. Aos vinte e sete anos sofreu um colapso nervoso. Aos vinte e nove anos foi derrotado ao se candidatar à presidência da Câmara. Aos trinta e um anos foi derrotado no Colégio Eleitoral. Aos trinta e nove anos foi derrotado na sua candidatura ao Congresso. Aos quarenta e seis anos foi derrotado para o Senado. Aos quarenta e sete anos foi derrotado na sua candidatura à vice-presidência. Aos quarenta e nove anos foi derrotado novamente na sua candidatura ao Senado. Até que aos cinqüenta e um anos foi eleito presidente dos Estados Unidos. Sem desmerecer os demais, Lincoln foi considerado o maior presidente que os Estados Unidos já tiveram, aquele que preservou a unidade da nação e foi seu grande líder durante a sangrenta Guerra de Secessão. Faça como Lincoln: não deixe que as derrotas do presente lhe roubem a esperança de obter vitórias no futuro. Você não foi criado para ser um fracasso. Sua vocação é a vitória. Seja um vencedor de gigantes!

4. Em vez de temer a ameaça, transforme-a no instrumento de sua mais consagradora vitória.

A espada de Golias era uma arma poderosíssima e assaz temida. Quem ousaria enfrentar aquele gigante cara a cara? Quem ousaria medir forças com aquele brutamontes? Mas Davi não só derrubou o arrogante gigante com uma pedrada, mas decepou a sua cabeça com a sua própria espada. A espada de Golias tornou-se o instrumento de sua própria ruína. Davi usou algo que o ameaçava para livrar-se de vez do ameaçador. Ele transformou o perigo que o ameaçava no golpe fatal para o próprio adversário.

Precisamos olhar para os problemas sob nova perspectiva. Temos que olhar para a vida pela ótica de Deus. Precisamos transformar vales em mananciais, desertos em pomares, vale da destruição em portas da esperança, a espada do gigante em símbolo da nossa retumbante vitória.

Og Mandino, um conhecido escritor, enfrentou uma grave crise pessoal aos trinta e cinco anos de idade: seus negócios faliram e sua esposa o abandonou. Começou a beber e tornou-se um alcoólatra, passando a viver nas sarjetas. Chegou a ponto de pensar em suicídio. Dez anos depois, no entanto, estava no auge da fama como escritor, com livros traduzidos e lidos no mundo inteiro. Ele sacudiu a poeira. Levantou-se das cinzas. Despojou-se do opróbrio, olhou para a frente, aprendeu com os fracassos do passado e fez deles a estrada para o sucesso. Seus livros buscam encorajar pessoas desalentadas a olhar para a vida com otimismo. Muitas pessoas têm encontrado ânimo nos seus escritos. Outros têm se desvencilhado das peias do passado para construir novos sonhos no futuro. Vencedores de gigantes não moram no passado, não vivem de reminiscências, não ficam presos no cipoal das memórias amargas. Eles olham para o futuro com esperança, não capitulam diante das crises, e transformam tragédias em triunfo.

Em sua autobiografia, Og Mandino oferece um conselho prático para as pessoas que desejam viver com otimismo. Comece o dia lendo o jornal da sua cidade, diz ele. Mas evite iniciar a leitura pelas páginas políticas, pois isso poderia deprimi-lo. Evite começar também pelas notícias econômicas, pois poderia deixá-lo desanimado. Também não é aconselhável iniciar pela leitura das páginas policiais, pois você poderia ficar com medo de sair de casa. Não é prudente começar o dia lendo as notícias esportivas, pois seu time pode ter sofrido uma goleada no último jogo. Assim, se você pretende começar o dia com entusiasmo, você deve iniciar a leitura pela página do obituário, que traz a lista das pessoas que morreram. É isso mesmo. Leia a lista toda, sem pular um nome sequer. Ao final da leitura, você vai descobrir algo fantástico: seu nome não consta da lista. As pessoas que estão nessa lista dariam tudo para estar no seu lugar, mas elas estão mortas, e você está vivo! E por estar vivo, hoje pode acontecer um milagre na sua vida. Não abra mão da vitória, você é um vencedor de gigantes!

7

Vencedores de gigantes dedicam suas vitórias a Deus

Disse mais Davi: O Senhor me livrou das garras do leão e das do urso; ele me livrará das mãos deste filisteu [...] Davi, porém, disse ao filisteu: Tu vens contra mim com espada, e com lança, e com escudo; eu, porém, vou contra ti em nome do Senhor dos Exércitos, o Deus dos exércitos de Israel, a quem tens afrontado. Hoje mesmo, o Senhor te entregará na minha mão; ferir-te-ei, tirar-te-ei a cabeça e os cadáveres do arraial dos filisteus darei hoje mesmo às aves dos céus e às bestas-feras da terra; e toda a terra saberá que há Deus em Israel. Saberá toda esta multidão que o Senhor salva, não com espada, nem com lança, porque do Senhor é a guerra, e ele vos entregará nas nossas mãos (1 Sm 17.37,45-47).

Vencedores de gigantes não são soberbos. A soberba é o portal da ruína, é a ante-sala do fracasso. Nenhum homem é verdadeiramente grande se não for humilde. Os nobres não se exaltam. Os vencedores depositam suas coroas diante daquele que está assentado no trono. A arrogância é a sepultura dos derrotados; a humildade, o pódio dos vencedores.

Davi nos ensina algumas lições preciosas:

1. O Deus que fez maravilhas ontem, pode intervir milagrosamente na nossa vida hoje.

Saul não acreditou no sucesso de Davi por julgá-lo muito jovem e inexperiente para tão grande desafio. Contudo, diante da incredulidade de Saul, Davi disse: *"O Senhor me livrou das garras do leão e das do urso; ele me livrará das mãos deste filisteu"* (1 Sm 17.37). O Deus que agiu no passado é o mesmo que age hoje. Deus não é uma relíquia, um objeto sagrado encostado numa prateleira de museu. Ele não é apenas o Deus dos antigos. Ele está vivo. Ele é imutável. Ele nunca abdicou do seu poder. Seu braço não se encolheu. Ele opera maravilhas ainda hoje.

Não podemos concordar com aqueles que tentam limitar o poder de Deus, dizendo que o tempo dos milagres cessou. Deus se recusa a ser trancado dentro dos estreitos conceitos teológicos de homens de mentes fechadas. Ele livrou ontem e livra hoje, salvou ontem e salva hoje. Ele curou os enfermos ontem e cura hoje, libertou os oprimidos ontem e liberta hoje. Foi nesta verdade que Davi se agarrou para enfrentar o gigante. Davi não era louco para desafiar Golias estribado em sua própria força. Ele sabia que a vitória vem de Deus, do Deus que nunca muda.

Não glorificam a Deus aqueles que, embora crendo na veracidade dos milagres passados, se recusam a crer na poderosa intervenção de Deus hoje. Quando Jesus chegou em Betânia, depois que Lázaro já estava sepultado havia quatro dias, Marta expressou toda a sua tristeza e decepção, dizendo: *"Senhor, se estiveras aqui, não teria morrido o meu irmão. Mas também sei que, mesmo agora, tudo quanto pedires a Deus, Deus to concederá"* (Jo 11.21,22). Então Jesus lhe declarou: *"Teu irmão há de ressurgir"* (Jo 11.23). Ela prontamente retrucou: *"Eu sei que ele há de ressurgir na ressurreição, no último dia"* (Jo 11.24). Mais tarde, quando Jesus mandou tirar a pedra do túmulo de Lázaro, Marta mais uma vez interferiu dizendo: *"Senhor, já cheira mal, porque já é de quatro dias"* (Jo 11.39). Então, Jesus lhe respondeu: *"Não te disse eu que, se creres, verás a glória de Deus?"* (Jo 11.40). Marta acreditava que Jesus poderia ter impedido a morte de seu irmão. Acreditava que Jesus iria ressuscitá-lo no último dia, mas não conseguia enxergar que Jesus poderia levantá-lo da morte naquele momento. Ela acreditava no Deus que agiu no passado e no Deus que agirá amanhã, mas tinha dificuldade para crer no Deus que age no presente.

2. A vitória é dádiva de Deus, não mérito do homem.

Davi entrou na peleja sabendo que Deus é quem iria conduzi-lo em triunfo. A vitória vem de Deus. É Ele quem adestra as nossas mãos para a batalha. Não vencemos por causa das armas nem por causa das nossas estratégias, mas por causa da intervenção divina. Davi disse para Golias: *"Hoje mesmo o Senhor te entregará na minha mão"* (1 Sm 17.46). A vitória sobre Golias não resultou da luta, mas da ação de Deus. É Ele que nos dá a vitória. Podemos nos sentir como aquele camundongo que atravessou uma velha ponte nas costas do elefante. No meio do trajeto, a ponte começou a balançar e o perigo de um acidente trágico parecia ser iminente. O desastre seria fatal. A morte, uma conseqüência inevitável. Mas, enfim, conseguiram alcançar o outro lado, salvos e seguros. O camundongo olhou para o elefante e disse: "Rapaz, como nós chacoalhamos aquela ponte, hein!" Quando andamos com Deus nos sentimos como um camundongo, mas com a força de um elefante; e depois de atravessarmos as águas turbulentas da vida, depois de cruzarmos os abismos perigosos da jornada, podemos dizer: "Deus, nós chacoalhamos aquela ponte, hein!"

Saul teve medo porque olhou para as circunstâncias, Davi triunfou porque olhou para Deus. Quem olha para os gigantes é dominado pelo pânico, mas quem olha para Deus e sabe que a vitória vem dele, enfrenta os gigantes e vence. Davi não lutou confiado em sua experiência nem em suas armas. Ele avançou contra o gigante em nome do Senhor dos Exércitos. Ele enfrentou o gigante não porque queria notoriedade, mas porque tinha zelo pelo nome de Deus.

3. Nossas vitórias devem encorajar outros a se tornarem vencedores.

Quando os soldados de Israel viram a coragem e a vitória de Davi, animaram-se e puseram-se a lutar com galhardia (1 Sm 17.52,53). Bastou que eles vissem um líder em posição de combate e disposto a vencer, para que fossem tomados por um novo ânimo. Precisamos de modelos. Precisamos de referências. Precisamos de homens que sirvam de inspiração para que outros saiam do comodismo. Precisamos de pessoas que mexam com os nossos brios e nos desafiem a sair do marasmo. O verdadeiro líder é aquele que desperta os outros a segui-lo. Não há líderes solitários. Os líderes começam sozinhos, mas nunca terminam sozinhos. Atrás deles há um exército que se levanta da mediocridade e empunha as armas da vitória. O líder é aquele que move e comove. Desperta pela palavra e pelo exemplo.

O líder é aquele que, através do seu heroísmo, tira os covardes da caverna do medo. Em vez de capitular frente ao desânimo da maioria, o líder desperta essa maioria para seguir as suas pegadas. Davi enfrentou problema semelhante em uma outra ocasião. Sua família, seus bens e as famílias de seus seiscentos homens foram saqueados na cidade de Ziclague (1 Sm 30.1-20). A cidade foi destruída e queimada. Os amalequitas deixaram um rastro de destruição atrás deles. Quando Davi e seus homens chegaram à cidade e a viram devastada e espoliada, com suas famílias e seus bens nas mãos do inimigo, eles *"choraram até não terem mais forças para chorar"* (1 Sm 30.1-4). Nesse momento de crise, os homens de

Davi voltaram-se contra ele, culpando-o pela tragédia, e pegaram em pedras para matá-lo. Davi estava só e sob oposição cerrada dos seus aliados. Ele já tinha sido atacado pelos inimigos de fora, e agora estava sendo atacado pelos de dentro de sua própria casa . No auge da angústia, Davi se voltou para Deus, reanimando-se no Senhor seu Deus, e pedindo a Ele que lhe desse discernimento e direção. Depois de orar, ele não apenas estava curado do seu desespero, mas tinha também o apoio dos seus homens e a promessa de vitória do próprio Deus. Davi lutou e venceu. Tomou de volta tudo o que o inimigo lhe roubara. Esse é o perfil de verdadeiro líder, aquele que é capaz de influenciar positivamente as pessoas que estão à sua volta.

Durante os jogos pára-olímpicos disputados entre crianças e adolescentes em Seattle, nos Estados Unidos, aconteceu um fato extraordinário. O estádio estava lotado. A multidão esperava pelo início da corrida que seria disputada por crianças excepcionais. Assim que foi dada a partida, uma menina de doze anos destacou-se do grupo e assumiu a liderança da corrida. O estádio inteiro a aplaudia, enquanto ela avançava, seguida bem de perto por outra corredora. De repente, ela percebeu que sua concorrente tropeçou e caiu. Para surpresa da multidão, ela parou de correr e foi ajudar a rival a se levantar. A multidão gritava: "Não pare. Não pare. Prossiga. Corra!" Mas ela tapou os ouvidos ao grito ensurdecedor da multidão. Enquanto ajudava a outra a se levantar os outros corredores se aproximaram, e então, todos se deram as mãos, cruzando juntos a linha de chegada. Nossas vitórias adquirem mais sabor e significado quando podemos levar conosco outras pessoas. Davi não venceu sozinho. Ele encorajou os soldados de Saul a serem co-participantes do seu triunfo. Você encoraja ou desestimula as pessoas à sua volta?

4. As honras pela vitória devem ser prestadas a Deus e não aos homens.

Davi creu que a vitória veio de Deus. Foi Deus quem entregou Golias nas mãos dele (1 Sm 17.46). Por isso é o nome de Deus e não o de Davi que deve ser honrado. *"Toda a terra saberá que há Deus em Israel"*, disse Davi (1 Sm 17.46). A vitória dos filhos de Deus torna o nome de Deus conhecido entre os povos. A vitória dos filhos de Deus é um testemunho entre as nações.

Mas a vitória dos filhos de Deus também produz impacto entre aqueles que têm medo de lutar, bem como na vida daqueles que nos menosprezam. Disse Davi: *"Saberá toda esta multidão que o Senhor salva, não com espada, nem com lança; porque do Senhor é a guerra, e ele vos entregará nas nossas mãos"*. Precisamos de vitórias para que o nome de Deus seja conhecido e também para que outros sejam encorajados. Precisamos de vitórias para fazer o nome de Deus notório aos nossos adversários, de sorte que as nações tremam na presença do Senhor (Is 64.2).

Não podemos nos alimentar apenas dos gloriosos feitos do passado nem nos reunir só para relembrarmos os portentosos feitos de ontem. A igreja precisa de vitórias hoje. Os soldados de Cristo precisam se colocar de pé hoje. Eles precisam de testemunhos vivos da ação de Deus hoje.

Davi reconheceu que a vitória veio de Deus e a glória devia ser tributada a Ele. Deus não reparte sua glória com ninguém. O único nome digno de ser proclamado, conhecido e temido em toda a terra é o nome do Senhor. O vaso é de barro, o instrumento é frágil, por isso a glória é de Deus. Mais tarde, quando as mulheres de Israel começaram a exaltar o nome de Davi, isso trouxe problemas para ele e para o reino, despertando inveja em Saul e provocando a mais insana perseguição contra Davi. Cometemos um grande equívoco ao promover o nome dos homens. Precisamos depositar nossas coroas, nossas vitórias, nossas conquistas aos pés do Senhor. Todo joelho deve dobrar-se diante dele. Só o seu nome é excelso, digno de toda honra, glória e louvor. O gigante caiu não por causa de Davi, mas por causa de Deus. Davi foi apenas o instrumento. E ele sabia disso.

O mais importante não é colecionar vitórias, mas ter intimidade com o Deus que nos dá a vitória. O mais importante não é erguer o troféu, mas depositar nossas coroas aos pés do Rei dos reis. O mais importante não é ser conhecido na terra, mas amado no céu, não é possuir as glórias deste mundo, mas possuir o Filho, o herdeiro de todas as coisas. Conta-se que um homem muito rico empregou toda a sua fortuna em quadros famosos. Esse homem tinha um único filho, que foi ferido mortalmente num combate. Um amigo mais achegado, tentando consolar o pai enlutado, pintou num quadro o rosto de seu filho amado. O pai ao receber o quadro, colocou-o em uma linda moldura. Anos depois, quando já estava prestes a morrer, aquele homem fez um testamento, deixando ordens para que seus quadros fossem leiloados e o dinheiro destinado às obras de caridade.

Assim foi feito, e todos aqueles quadros famosos foram expostos para serem vendidos em leilão. Mas para surpresa das pessoas presentes ao leilão, o primeiro quadro a ser anunciado foi aquele que retratava o filho. Não havia beleza naquele quadro. Nada que pudesse atrair o gosto de pessoas tão refinadas. Ninguém demonstrou interesse pelo quadro, e com muito custo, alguém o arrematou por mísera quantia. Para espanto geral, assim que o quadro foi vendido, o leiloeiro oficial anunciou o encerramento do leilão, explicando: "O meu senhor deixou estipulado no testamento que aquele que adquirisse o quadro do filho, teria todos os outros quadros". Quem tem o Filho tem tudo. A Bíblia diz que aquele que tem o Filho tem a vida. Deus é melhor do que suas próprias dávidas. Ele é mais excelente do que as mais excelentes vitórias que colecionamos na vida. Deus é a causa, o agente e a finalidade da nossa vitória. Dele, por meio dele e para Ele são todas as coisas. A Ele seja a glória eternamente. Amém.

Conclusão

A batalha contra os gigantes é uma peleja contínua. Quando um cai, outros se levantam. Não podemos ensarilhar as armas nem ficar desatentos. A vida cristã não é uma colônia de férias, mas um campo de batalha. Você é um soldado e foi alistado no exército de Deus. Você já recebeu as armas de defesa e combate. O poder para a vitória vem de Deus. O segredo da vitória é andar na dependência de Deus.

É bem verdade que os inimigos são muitos, e empregam vários métodos e estratégias. Eles são perseverantes. Precisamos estar atentos. Dormir na peleja pode ser um erro fatal. Deixar de vigiar pode acarretar grandes baixas. Depois de grandes vitórias, precisamos continuar firmes, pois outros ataques virão. Essa guerra continuará até que Jesus coloque todos os inimigos debaixo dos seus pés.

Mas você não precisa temer. O problema maior não é a presença do inimigo, mas a ausência de Jesus. Com Jesus ao seu lado você é mais do que vencedor. Se Deus está com você, ninguém poderá lhe resistir. Sua vocação é a vitória. Você foi marcado para vencer. Você está alistado nas fileiras daquele que já venceu, continua vencendo e consumará sua vitória final e retumbante quando o sistema do mundo, o anticristo, o falso profeta, o diabo, a morte e todos os ímpios forem lançados no lago de fogo.

As circunstâncias podem dar a impressão que o inimigo está prevalecendo. O diabo está fazendo grandes estragos nas famílias, nas instituições, nas nações. O mundo está gemendo de dores. A terra está em agonia aguardando a sua redenção. Filosofias ateístas, que negam a Deus e zombam da sua Palavra são trombeteadas nas cátedras eletrônicas, proclamadas nos centros universitários e despejadas nas mentes das crianças desde o início da lida estudantil. A corrupção dos valores, a delinqüência galopante, a paganização da cultura, a desmoralização da família se agigantam e sentimo-nos como que esmagados debaixo das botas desses gigantes. Mas, nesses tempos de trevas, precisamos levantar os olhos e contemplar nosso Deus assentado sobre um alto e sublime trono. A história não caminha para um fim trágico. Jesus está assentado na sala de comando do universo. Ele é digno para conduzir a história à consumação final. A vitória é certa e segura. Caminhamos para a coroação. Por enquanto o inimigo revela a sua insolência, mas sua derrota final já está lavrada. Levante a cabeça, entre em campo, lute e vença.

No dia primeiro de janeiro de 1919, o time de futebol da Universidade da Califórnia disputava o final do campeonato com o time da Universidade da Geórgia. O estádio estava lotado. Uma multidão aguardava ansiosa para ver o grande ídolo Roy Rigels entrar em campo para dar a vitória ao seu time. Ele era a esperança de vitória. Mas no primeiro tempo, ele frustou a multidão com um péssimo desempenho e até mesmo com um gol contra. Saiu de campo vaiado, envergonhado e com o estigma do fracasso. No vestiário não conseguiu esconder suas lágrimas nem sua desesperança. Parecia que o mundo havia desabado sobre sua cabeça. A derrota parecia certa e inevitável.

Nesse momento, o técnico se aproximou, passou a mão em sua cabeça e lhe disse: "Coragem, o jogo ainda não acabou. Você ainda tem o segundo tempo. Volte ao campo, lute e vença". Roy Rigels levantou a cabeça, voltou ao campo, e jogou como nunca antes havia jogado em toda sua vida. Ele conseguiu virar o placar e conquistou uma retumbante vitória, deixando o campo sob os aplausos da delirante multidão.

Caro leitor, o jogo ainda não acabou. A luta continua, mas a vitória é certa. Você pode ser um vencedor de gigantes!

Esta obra foi composta pela
Atis Design Ltda
e impressa pela
Imprensa da Fé
na fonte Adobe Garamond corpo 11,
sobre Off-set 75 g/m² no miolo e
capa Cartão Supremo 250 g/m² .
São Paulo, Brasil, Verão de 2005.